Managing Open Service Innovation

Open Innovation: Bridging Theory and Practice Volume 6

Managing Open Service Innovation

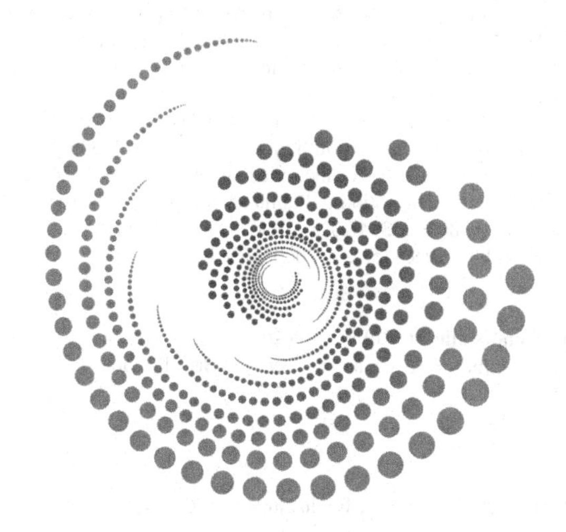

Editors

Anne-Laure Mention
RMIT University, Australia

Tor Helge Aas
University of Agder, Norway

NEW JERSEY • LONDON • SINGAPORE • BEIJING • SHANGHAI • HONG KONG • TAIPEI • CHENNAI • TOKYO

Published by

World Scientific Publishing Co. Pte. Ltd.
5 Toh Tuck Link, Singapore 596224
USA office: 27 Warren Street, Suite 401-402, Hackensack, NJ 07601
UK office: 57 Shelton Street, Covent Garden, London WC2H 9HE

Library of Congress Cataloging-in-Publication Data
Names: Mention, Anne-Laure, editor. | Aas, Tor Helge, editor.
Title: Managing open service innovation / editors, Anne-Laure Mention, Royal Melbourne Institute of Technology, Australia, Tor Helge Aas, University of Agder, Norway.
Description: Hackensack, New Jersey : World Scientific, [2021] | Series: Open innovation : bridging theory and practice, 2424-8231 ; volume 6 | Includes bibliographical references and index.
Identifiers: LCCN 2020056141 | ISBN 9789811234484 (hardcover) | ISBN 9789811234491 (ebook for institutions) | ISBN 9789811234507 (ebook for individuals)
Subjects: LCSH: Service industries--Technological innovations. | Service industries--Information technology. | Diffusion of innovations. | Public administration--Information technology.
Classification: LCC HD9980.5 .M344 2021 | DDC 658.4/063--dc23
LC record available at https://lccn.loc.gov/2020056141

British Library Cataloguing-in-Publication Data
A catalogue record for this book is available from the British Library.

For any available supplementary material, please visit
https://www.worldscientific.com/worldscibooks/10.1142/12214#t=suppl

Desk Editors: Balamurugan Rajendran/Daniele Lee

Typeset by Stallion Press
Email: enquiries@stallionpress.com

Printed in Singapore

https://doi.org/10.1142/9789811234491_fmatter

About the Editors

Anne-Laure Mention is the Director of the Global Business Innovation Enabling Capability Platform at RMIT, Melbourne, Australia. She is also a Professor at the School of Management at RMIT, Melbourne; a Visiting Professor at Université de Liège, Belgium and, a Visiting Professor at Tampere University of Technology, Finland and a Fintech and Blockchain Visiting Fellow at Singapore University of Social Sciences. She holds several other visiting positions in Europe and Asia. Anne-Laure is one of the founding editors of the *Journal of Innovation Management*, and was the Deputy Head of the ISPIM Advisory Board (2012–2018). She is the co-editor of a book series on *Open Innovation*, published by World Scientific/Imperial College Press. Her research interests revolve around open and collaborative innovation, innovation in business to business services, with a particular focus on financial industry and FinTech, technology management, and business venturing. She has been awarded twice the prestigious IBM Faculty Award for her research on innovation.

Tor Helge Aas serves as an Professor at University of Agder and Research Professor at NORCE Norwegian Research Centre. He holds a PhD in strategy and management from Norwegian School of Economics and a MSc in Information and Communication Technology Management from University of Agder. Tor Helge Aas is the Leader of the Research Group Strategy, Innovation and Entrepreneurship at University of Agder. He is researching innovation management, and his research concentrates on topics such as the organizational effects of innovation, innovation processes and capabilities, collaboration for innovation, and management control of innovation. The research conducted by Tor Helge Aas has been published in international journals such as *Technovation*, *International Journal of Innovation Management*, *Service Industries Journal and Journal of Service Theory and Practice* among others. He has also co-authored book chapters in books published by, among others, Routledge and Cambridge Scholars Publishing.

https://doi.org/10.1142/9789811234491_fmatter

About the Contributors

Pierre-Jean Barlatier, PhD Habil. is an Associate Professor of Strategy at EDHEC Business School. His research focuses on strategic management and organization design applied to open innovation management. His works has recently been published in *Research Policy, Technological Forecasting and Social Change* and *Strategic Organization* among others. He is currently member of the board of the *Association Internationale de Management Stratégique* (AIMS), the main French-speaking scientific association in Strategy and Organization Management. Pierre-Jean is also an Associate Researcher at BETA CNRS-University of Strasbourg (France) and has been appointed to several Visiting Researcher positions at the Royal Melbourne Institute of Technology (RMIT) University (Australia), the University of Technology Sydney (UTS) Business School (Australia) as well as Marie Curie Fellow at Copenhagen Business School (CBS, Denmark). He holds an HDR from the University of Strasbourg (France, 2015) and a PhD in Management Sciences from the University of Nice-Sophia Antipolis (France, 2006). Prior to joining EDHEC, he was Researcher at the Luxembourg Institute of Science and Technology (LIST, Grand-Duchy of Luxembourg).

Mohammad Ejaz is an Associate Professor in marketing management at University of South-Eastern Norway. He obtained his PhD in business and economics in 2016 and joined the USN School of Business in 2017. His research interests span the new service development, open innovation, and innovation in internet-based aggregation services.

Niels Frederik Garmann-Johnsen is a Senior Researcher and Project Manager at NORCE — Norwegian Research Centre, and Associate Professor at the Department of Information Systems at the University of Agder. He holds a PhD in social sciences with a specialization in information systems. Information systems means the totality and interaction between ICT, processes, and people, not just information and communication technology. Garmann-Johnsen belongs to UiA's Center for Digital Transformation, CEDIT, and it is also affiliated with UiA's Center for e-Health — research forum, as well as the ERCIS network and its Competence Center for Digital Transformation in SMEs. Garmann-Johnson's doctoral dissertation is about collaborative processes and prerequisites for succeeding with eHealth innovation based on the coordination reform in healthcare. Garmann-Johnsen belongs to the research group Work Life and Innovation in NORCE research, dept. for Society.

Eleni Giannopoulou is a Statistical Officer at the European Commission. Prior to that she has worked as a Researcher for the Luxembourg Institute of Technology in Luxembourg. Eleni Giannopoulou has obtained a diploma (Msc) in electrical and computer engineering from the National Technical University of Athens (Greece), a master's degree in management and economics of innovation from Chalmers University of Technology (Sweden) and a PhD in economics from the University of Strasbourg (France). Her research interests focus on economics of innovation and technological transformation, academia-industry collaboration for innovation, innovation intermediaries, technology management, strategic management of innovation and open innovation. Her work has been published in academic journals such as *Research Policy*, *Managing Service Quality*, *International Journal of Innovation Management*, and *Journal of Technology Management and Innovation* among others.

Julius Francis Gomes is pursuing his PhD in international business from the University of Oulu. He currently works at the Oulu Business School as a doctoral student to research the futuristic business models for digital intensive industries. His research focuses on using business models as a mean to look into future industries & ecosystems. He is interested to research business ecosystems in different contexts like cyber security, healthcare, future's network etc. with a business model perspective. He received his MSc (2015) in international business from the University of Oulu. Prior to that he acquired MBA (2011) specializing in managing information systems in business applications. Previously, he has also enjoyed about three years in a top tier bank in Bangladesh as a channel innovator.

Lidia Gryszkiewicz is an Innovation and Technology Management Expert and an Experienced Strategist. She has completed the Creating Shared Value programme at Harvard Business School and holds a PhD in innovation management. She worked as a strategy consultant for Arthur D. Little and run strategic projects for PWC Amsterdam and carried out research as R&D engineer for Luxembourg Institute of Science & Technology. After co-founding Limitless (www.limitless.lu), she has co-founded The Impact Lab (www.theimpactlab.org), a collaborative innovation strategy consultancy advising Ministers and Secretaries of State, city mayors, businesses and non-profits. Her work has been featured in Stanford Social Innovation Review, Forbes, scientific journals and international conferences. She is the European Commission's expert on technology, innovation, and entrepreneurship, serves as a mentor at Founder Institute and is an advisor of the World Economic Forum's Global Shapers Community.

Juha Häikiö is a Research Scientist at the VTT Technical Research Centre of Finland and works in the Foresight-driven Business Strategies unit. He holds an MSc in Information Processing Science. His research interests include user-centered design, user experience and digital service ecosystems. He has experience about R&D projects focusing on digitalization in a number of different industrial sectors.

Kirsti Mathiesen Hjemdahl is a PhD in Art from the University of Bergen, within Ethnology. From the position as head of research at Agderforskning, she has been the research leader of the INSITE project which is about data driven decision processes. Hjemdahl became Professor II at the University of Agder in 2019, as well as head of research within innovation at NORCE Research Centre. She has published within the fields of popular culture, political transformations, applied research and innovation within tourism, culture and creative industries. Presently, Hjemdahl is CEO at the Cultiva Foundation in Norway.

Katja Maria Hydle is the Head of Research at NORCE Research and holds a PhD in Strategic Management from BI Norwegian Business School. She graduated in Political Science from University of Lausanne, Switzerland, holds a MA in advanced European studies from College of Europe, Belgium. Her research concentrates on innovation, strategy, professional service work, organizational practices and multinational companies, with outlets in leading academic journals such as *Organization Studies*, *Human Relations*, and *Journal of Word Business*.

Marika Iivari is currently working as a Business Development Executive at Fingersoft Ltd. She has a PhD at the AACSB accredited Oulu Business School, Finland. She holds a MSc in international business degree from the Ulster University, Northern Ireland. She is completing her doctoral dissertation on business models and open innovation in the context of innovation ecosystems. She is currently working on a research project on Industrial Internet. Previously she has worked on living labs and open innovation ecosystem in the context of integrative urban planning and smart cities.

Christina Lee is an Education Professional working in learning design for a leading online education provider. She has been a research assistant at RMIT University, School of Management, where she researched open service innovation. Her specialty areas are education, digital innovation and innovation in learning. Christina has used her knowledge of innovation to co-develop and introduce game design, programming and animation to secondary schools Australia wide.

Santiago Martinez research focuses on user-centred design and usability. Martinez background is Computing Engineering and he has been working in Human–Computer Interaction (HCI) for the last 9 years. In 2009, Martinez was awarded the Alison Armstrong Research scholarship for his PhD in HCI in the University of Abertay, UK, where he worked in an interdisciplinary environment with Psychologists, Sociologists and Health professionals. He is focused on how to involve users without Information and Telecommunication Technology (ICT) experience, such as the occasional, first-time, and disabled, in the design and evaluation of solutions for them. Martinez have developed specific methodologies for his research and worked with a wide range of the established ones, such as Participatory Design, User-Centred Design and Ethnography. In the private sector, Martinez has experience in working in the Health Public Sector in Spain in a multi-strategy IT consulting company (Everis Spain Ltd.).

Daniel Nordgård is an Associate Professor at the University of Agder and Senior Researcher at NORCE. He has a broad background from music, foremost as musician and artist. His research is very much devoted to the music- and the cultural industries, with a special emphasis on digital change. In 2013, Nordgård was appointed by the Norwegian Government to lead their committee on music streaming and digital change in the music industry and wrote the report from the committee. He was also member of the COST-financed network: Dynamics of Virtual Work, looking at digitalization and work. He holds several positions in different boards, nationally and internationally, including Gramo (Norwegian collecting society for recording artists and record companies), The Norwegian Film Institute, Gramart (The Norwegian featured artist organization). He also sits on the board of the International Music Business Research Association (IMBRA).

Minna Pikkarainen is a Joint Connected Health Professor of VTT Technical Research Centre of Finland and University of Oulu/Oulu Business School, Martti Ahtisaari Institute and Faculty of Medicine. As a Professor of connected health Minna is doing on multidisciplinary research on innovation management, service networks and business models in the context of connected health service co-creation. Professor Pikkarainen has extensive record of external funding, her research has been published in large number of journals and conferences, e.g., in the field of innovation management, software engineering and information systems. During 2006–2012, Professor Minna Pikkarainen has been working as a Researcher in Lero, the Irish software engineering research centre, researcher in Sirris, collective "centre of the Belgian technological industry" and business developer in Institute Mines Telecom, Paris and EIT (European Innovation Technology) network in Paris and Helsinki. Her key focus areas as a business developer has been in healthcare organizations.

Rómulo Pinheiro is a Professor of Public Policy and Administration at the University of Agder (UiA), Norway, where he co-heads the research group on Public Governance and Leadership (GOLEP). Rómulo's research interests are located at the intersection of public policy and administration, organizational theory, economic geography, innovation and higher education studies. His work has been published in several scientific journals, such as, *Public Administration Review*, *Science and Public Policy*, *Higher Education*, *Studies in Higher Education*, *European Journal of Higher Education*, *Cambridge Journal of Regions, Economy and Society*, *Scandinavian Journal of Public Administration*, *Tertiary Education and Management*, *City, Culture and Society*, *International Journal of Cultural Policy*, etc. Rómulo has co-edited numerous books by Springer, Routledge, Palgrave, Sense, Emerald. He has secured close to EUR 3 million in external, competitive funding, and is currently leading a number of comparative projects.

Teemu Santonen received his PhD (Econ.) degree in information systems science from Aalto University in Finland in 2005. Currently he is acting as a Lecturer for "Knowledge Intensive Business Services (KIBS)" at the Laurea University of Applied Sciences in Finland. Prior to his academic career Santonen was working over a decade as a consult and a development manager in leading Finnish financial, media and ICT sector organizations. Santonen is also a scientific panel member of ISPIM and is former board member of Finnish Strategic Management Society. Currently his research interest focuses on "Social Network Analysis (SNA)", "Scientometrics", "Innovation management" and "living labs". In Laurea, Santonen has also filed several invention disclosures, which have resulted a start-up company and one patent. The Finnish Inventor Support Association have honored Santonen's novel crowdsourcing project as the best school related innovation in Finland.

Kristin Wallevik is Dean of the School of Business and Law at the University of Agder. She holds a PhD-degree from Copenhagen Business School and an MBA from The Monterey Institute of International Studies. Her current research areas are innovation management, competitiveness, regional innovation systems and industry development. She has published chapters, reports and articles in Routledge, Palgrave Macmillan and European Planning Studies, to mention some. She has published several scientific reports and held speeches nationally and internationally on topics like industry development, innovation, sustainability and cluster development.

Erik Wästlund received his PhD in psychology from Göterbors universitet 2007. Wästlunds research is focused on the intersection between technology and human behavior. His research has encompassed digitalization in healthcare and retail as well as hospitality.

Peter Ylén is a Principal Scientist in the business, innovation and foresight research area at VTT in Espoo, Finland. He has worked at VTT since 2005 and before that as a managing director and research director in the private sector. He has also worked as a Researcher and Professor at Aalto University and University of Vaasa respectively. He is active in several scientific associations and board member in various companies. His scientific interests are in system theory, service science, business ecosystems, modelling and simulation. He is the owner of several EU and TEKES funded projects and has led "System dynamics and optimization" research team and "Business ecosystem, value chain and foresight" research area at VTT.

https://doi.org/10.1142/9789811234491_fmatter

Contents

https://doi.org/10.1142/9789811234491_0001

Chapter 1

Managing Open Service Innovation: An Introduction

Tor Helge Aas[*,†,§] *and Anne-Laure Mention*[‡,¶]

[*]*School of Business and Law, University of Agder, Campus Kristiansand, Universitetsveien 25, 4630 Kristiansand, Norway*

[†]*NORCE Norwegian Research Centre, Nygårdsgaten 112, 5008 Bergen, Norway*

[‡]*RMIT University, 124 La Trobe St, Melbourne VIC 3000, Australia*

[§]*tor.h.aas@uia.no*

[¶]*anne-laure.mention@rmit.edu.au*

Abstract. Two of the hottest topics in the area of innovation management in the last decade have been service innovation and open innovation. The two phenomena are often studied in isolation. It has, however, been suggested that service innovation processes are often characterized by more openness than closeness and that more research is needed to understand how organizations manage open service innovation activities in different contexts. This chapter develops a framework for the management of open service innovation. The framework consists of four distinct contexts within the area of open service innovation management: (1) the management of inbound open service innovation processes, (2) the management of outbound open service innovation processes, (3) the management of resources needed in inbound open service innovation processes, and (4) the management of resources needed in outbound open service innovation processes.

The chapter also positions the remaining chapters of this book according to the framework and provides an overview of all chapters.

Keywords. Open innovation; service innovation; open service innovation management; conceptual framework; innovation processes; innovation resources; inbound open innovation; outbound open innovation.

1. The Motivation for the Book

Innovation is about transforming ideas into new or improved products, services or processes, to advance in the marketplace (Baregheh *et al.*, 2009), and two of the hottest topics in the area of innovation management in the last decade have been service innovation (Carlborg *et al.*, 2014) and open innovation (Bogers *et al.*, 2017). Service innovation refers to the development of new services by both service and manufacturing organizations in for-profit, as well as non-profit sectors, while open innovation refers to the implementation of innovation processes where organizations use knowledge from or share knowledge with external actors. Even with several studies now been conducted to provide insights on these phenomena, they still continue to draw considerable attention from researchers, managers and policymakers. Most studies, however, study the two phenomena in isolation, even if it has been suggested that the implementation of open innovation practices may be feasible when organizations develop new services (Chesbrough, 2011). Both from managerial and policymaker perspectives, more research is therefore needed to understand how organizations manage open service innovation activities in different contexts (Randhawa *et al.*, 2016). This book responds to this call for research by presenting 10 chapters (including this one) that provide novel research-based insights on how open service innovation activities are managed.

This introductory chapter is structured in the following manner: In the next two sections we review the literatures on open innovation and service innovation. Thereafter, we discuss the concept of open service innovation and develop a framework for the management of open service innovation activities. Finally, we give a brief overview of the chapters in the book and we position the different contributions in the framework.

2. Open Innovation

The term *open innovation* was coined by Chesbrough (2003), and one of the most cited definitions suggests that *open innovation is the use of purposive inflows and outflows of knowledge to accelerate internal innovation, and to expand the markets for external use of innovation, respectively* (Chesbrough *et al.*, 2006, p. 1). What resources organizations need to be able to innovate successfully has always been one of the main topics in innovation management research, but by introducing the concept of open innovation Chesbrough (2003) shifted the focus from organizations' internal resources to resources outside the borders of the organizations.

It should be noted here that Chesbrough (2003) was certainly not the first to argue that the use of external resources could be valuable for organizations. Chesbrough's (2003) ideas resonated well with other trends, for example, related to outsourcing and networking, in the management discipline in the late 1990s and early 2000s (Pittaway *et al.*, 2004), and researchers within the discipline of innovation management had already stated that *indeed, the locus of innovation is no longer the individual or the firm, but increasingly the network in which a firm is embedded* (Powell *et al.*, 1996). In fact, almost 20 years prior to Chesbrough (2003), Lundvall (1985) argued that the flow of knowledge among people, organizations and institutions is a key to successful innovation.

However, by coining the term "open innovation", Chesbrough (2003) created a common label to these ideas, and the new concept initiated a comprehensive stream of research on how knowledge flows between different actors during innovation processes (Huizingh, 2011). The shift from a focus on internal resources to a focus on external resources has proved to be very attractive also for managers, and gradually, the idea that organizations can and should innovate on their own in isolation has become outdated (Gassmann, 2006). In addition, the development of new information and communication technologies, such as the Internet, the last decades have made it possible for organizations to work more easily with each other, and in that way the development in the information and communication technology area has acted as a catalyst for the implementation of open innovation practices (Dodgson *et al.*, 2006).

Several reviews of the research literature on open innovation have been published. One example is the study by Dahlander and Gann (2010) who concluded that there are four types of open innovation. The first relates to inbound pecuniary knowledge flows and the second relates to inbound non-pecuniary knowledge flows, whereas the third and fourth type relate to outbound pecuniary and outbound non-pecuniary knowledge flows. According to the review by Dahlander and Gann (2010), the different types of open innovation have very different advantages and disadvantages. This illustrates how complex the concept of open innovation is from a management perspective. Open innovation is not one concept that is ready to implement for managers. Instead, a range of different variations exist where all variations are not suitable for all organizations and all contexts. Thus, an organization aiming to benefit from open innovation needs to be very aware of what type of advantages it would like to achieve and what kind of disadvantages it would like to avoid.

More recent reviews of the open innovation literature include, for example, West and Bogers (2014), Randhawa *et al.* (2016), and Bogers *et al.* (2017). These reviews found that the literature in this area has become more mature recent years, but that there are still knowledge gaps that hinder efficient implementation of open innovation practices in many contexts. West and Bogers (2014) particularly reviewed the literature discussing inbound open innovation and suggested a four-phase model in which a linear process of obtaining external knowledge, integrating the knowledge in the organization, and commercializing innovations, was combined with interaction between the organization and its collaborators. The review concluded that research on all these phases exist, but that *researchers have front-loaded their examination of the first phase in the process*

(West & Bogers, 2014, p. 814), that of obtaining knowledge, while our knowledge related to the other phases are more immature.

Furthermore, the study by Bogers *et al.* (2017) acknowledged that *a number of industry-level contingencies are relevant in the context of open innovation, also for explaining the effectiveness of open innovation across different settings* (p. 20), and Randhawa *et al.* (2016) particularly highlight that *service-related aspects of open innovation has received limited research attention, despite being regarded as highly pertinent to today's networked and service-led environment* (p. 767), and that open innovation researchers need to *enhance service focus and conceptualize "open service innovation"* (p. 764).

3. Service Innovation

The growing attention to service innovation is in part driven by the fact that in most developed countries the service sectors now have more employees and a larger share of the Gross National Product (GNP) than the manufacturing sectors (Gallouj & Djellal, 2010). At the same time, manufacturing firms are to an increasingly extent offering services either in addition to products or as a replacement of products (Baines *et al.*, 2017). These services are sometimes called "hidden services", because *in statistics they are registered as part of manufacturing's contribution to GNP* (Grönroos, 2007, p. 3). Thus, service innovation is playing an increasingly important role in developing the performance of organizations across both manufacturing and service sectors (Carlborg *et al.*, 2014).

A service innovation may simply be defined as *an offering not previously available to a firm's customers resulting from the addition of a service offering or changes in the service concept* (Menor *et al.*, 2002, p. 138), or more specifically, *as a new or considerably changed service concept, client interaction channel, service delivery system or technological concept that individually, but most likely in combination, that leads to one or more (re)new(ed) service functions that are new to the firm and do change the service/good offered on the market and do require structurally new technological, human or organizational capabilities of the service organization* (van Ark *et al.*, 2003, p. 16).

Researchers started to explore service innovation in the late 1980s, and Barras (1986) paper has been recognized as the first research paper discussing this type of innovation (Carlborg *et al.*, 2014). Most early service innovation research adopted a demarcation perspective. Pioneers in the field argued that service innovation should be a research area separated from the product innovation area, due to the fact that the specific characteristics of services, such as their intangibility, simultaneity, heterogeneity, and perishability, affect service innovation processes and make them different from product innovation processes (e.g., Edvardsson & Olsson, 1996; Sundbo, 1997).

Many researchers still apply this perspective, but today an increasing number of researchers instead apply a more unifying perspective, often referred to as the synthesis perspective, that acknowledges that service and product innovation also have many similarities (Gallouj & Windrum, 2009). Researchers applying this perspective also acknowledge a contingent view on innovation practices (Kuester *et al.*, 2013), but argue that different types of innovation should be studied and analyzed together through an integrated approach to better understand how organizations innovate (Hipp & Grupp, 2005). This approach *provides opportunities to better understand customer needs and value creation processes through combinations of services and products* (Carlborg *et al.*, 2014, p. 385).

In recent years, the research field of service innovation has matured and broadened, and a number of topics have received attention. Some researchers have been interested in the relationship between strategy and service innovation (Hydle *et al.*, 2014), and how organizations should be configured (Drejer, 2004). The capabilities and resources organizations need to build and possess to be able to carry out service innovation in a successful manner have been an important issue in this stream of research (Den Hertog *et al.*, 2010). Some researchers contributing to this literature have suggested that organizations often need to collaborate with external actors and utilize external resources when they develop new services (e.g., Aas, 2016), but most research has arguably focused on the importance of the resources and capabilities within the border of the organization (Dotzel *et al.*, 2013; Snyder *et al.*, 2016).

Other researchers have been more interested in the service innovation processes (Hipp & Grupp, 2005), including the organizational changes (Perks & Riihela, 2004) and business model changes (Aas & Jørgensen, 2016) that often happen in parallel with the new service development processes. One important question addressed in this stream of research has been how the service innovation processes are managed, and our knowledge about this topic is maturing (Aas *et al.*, 2017). Extant research has, however, only to a limited degree been able to reveal how complex open service innovation processes, that involve collaboration between different actors in the service ecosystem, are managed (Mina *et al.*, 2014).

4. Managing Open Service Innovation — A Framework

Management may be defined as *the process of organizing resources and directing activities for the purpose of organizational objectives* (Mercant & Van der Stede, 2013) and service innovation management frameworks, such as the framework suggested by Frohle and Roth (2007), often consist of two overarching dimensions that are relevant from a closed and from an open innovation perspective: (1) the management of innovation processes on the one hand, and (2) the management of innovation resources on the other hand. The first dimension relates to how both open and closed service innovation processes are managed, and the second dimension relates to the internal and external resources that are utilized in these processes, and how they are managed.

Most definitions and frameworks of the "open innovation" concept acknowledge that there are (at least) two types of knowledge flows in open innovation processes: (1) an inbound flow of knowledge from external actors that accelerate innovation in an organization, and (2) an outbound flow of knowledge from an organization

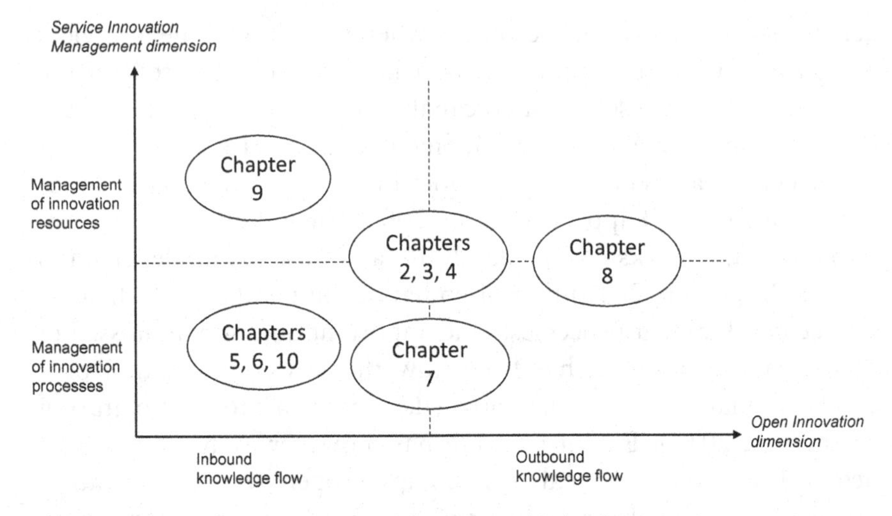

Figure 1. Chapters in this book.

that accelerate innovation in another organization (e.g., Dahlander & Gann, 2010).

If we combine the two dominating frameworks in the areas of service innovation management and open innovation, four dimensions of *open service innovation management* emerge: (1) the management of inbound open service innovation processes, (2) the management of outbound open service innovation processes, (3) the management of resources needed in inbound open service innovation processes, and (4) the management of resources needed in outbound open service innovation processes. These four dimensions represent four distinct research areas within the area of *open service innovation management*, and the framework is therefore well suited to position the remaining chapters in this book (see Figure 1).

5. Overview of the Chapters in this Book

Chapter 2

The second chapter is authored by Lee, Aas, and Mention and entitled "Open Service Innovation: A Systematic Literature Review".

Drawing on the logic of that open service innovation is closely intertwined with open innovation, they progress the discussion forward by scanning the state-of-the-art of open service innovation. A systematic literature review of peer-reviewed articles published in Scopus database between 2003 and 2018 is presented. Complementing the recent work by Randhawa *et al.* (2016), their results highlight that there is still no common place for open service innovation literature with papers found in 69 different journals, with majority of the studies being exploratory (qualitative or conceptual). Drawing on the idiosyncrasies between open innovation and service innovation, the chapter discusses the relevance of service dominant (SD) logic in an increasingly open and collaborative service environment. Authors identify a need for new theoretical development to unfold and progress open service innovation research and call for development of a firm-centric open dominant (OD) theory.

Chapter 3

The third chapter is authored by Santonen and entitled "Is there a Need for 'Open Service Innovation' Term: A Scientometrics Analysis of Open Service Innovation Research Domain". Santonen builds on Santonen and Conn (2015) to comprehensively examine open innovation and service innovation research trajectories. Through triangulation of published research, this chapter takes the reader on a journey of scientometric studies, exploring the perspectives (popularity, impact, network), levels of analysis (micro, meso, macro) and techniques (descriptive, co-citation analysis, social network analysis). The chapter concludes that a need for open service innovation term is redundant and draws attention towards scholarly cooperation in advancing research in the area.

Chapter 4

The fourth chapter is authored by Garmann-Johnsen and Martinez and entitled "Business Models for Collaborative eHealth in Homecare". Taking the discussions for conceptualization of open service innovation to its use in practice, this chapter presents a roadmap for service design and architecture development in the e-Health space. The chapter draws on e-business models and e-Health literature, incorporating a theoretical framework exploring requirements for enterprise systems and preconditions of efficiency and quality in e-Health adoption. The authors conclude with a call for further exploration examining the interactions between innovations and information structures based on the propositions presented in the paper.

Chapter 5

The fifth chapter is authored by Ejaz and Pinheiro and entitled "Innovation in Higher Education: From Contributor to Driver of Internet-Based Service Innovation". Continuing with the applications approach, this chapter captures the ever-increasing focus on the interaction between technology and educational environments. Authors highlight the role of universities in innovation processes and focus discussions on the journey of a new program at a university in Oslo, termed "virtual mobility". Contrary to previous research, in this case, authors find a lack of user integration in the innovation process and that higher education institutions seldom utilize the depth and breadth of their knowledge partners. A thought-provoking conclusion emerges — how to predict the value of open service innovations in a traditional setting, such as a classroom-based university environment? The authors suggest a need for longitudinal accounts of innovation processes at public universities.

Chapter 6

The sixth chapter is authored by Hydle and Wallevik and entitled "Intra-bound Innovation and Strategizing in Service MNCs". This chapter combines open innovation and strategy literature and exposes the open service innovation practices of 10 multinational companies. Authors present the geography of firm's innovation strategizing practices on a 2 × 2 dimension (explicit/emergent × exploitation/exploration) to identify the push and pull mechanisms at play. Interestingly, the chapter presents an alternate view to the role of MNC headquarters in driving innovation and pave the path for future research on "intra-bound innovation".

Chapter 7

The seventh chapter is authored by Barlatier, Giannopoulou, and Gryszkiewicz and entitled "New Service Development Process: What Can We Learn from Research and Technology Organizations?" Exploring the new service development (NSD) process of five renowned European research and technology organizations (RTOs), this chapter introduces the stage-wise NSD in RTOs and immerses the reader in the discussions on project-service dilemma. Highlighting the pre-NSD and post-NSD stages, authors draw attention to the value of research translation from academic desk to market. Several new avenues for research in the open service innovation area are proposed with focus afforded towards dilemmas, culture and role of RTOs in the innovation ecosystem.

Chapter 8

The eighth chapter is authored by Aas, Hjemdahl, Nordgård, and Wästlund and entitled "Outbound Open Innovation in Tourism: Lessons from an Innovation Project in Norway". Authors contribute towards ongoing discussions on outbound innovation in tourism by highlighting the role of knowledge mediators in the extent and type of knowledge transferred during bi-lateral meetings. The authors reveal that a degree of trust is deemed important in the knowledge transfer, with its sources and formative indicators being the call for future research.

Chapter 9

The ninth chapter is authored by Aas and entitled "Opening Up the Service Innovation Process Towards Ordinary Employees in Large Service Firms". Taking an internal view to open service innovation, author in this chapter focuses on the employees in a service firm. Opening up for open service innovation is found to be associated with managerial support, transparent systems, collaborative environment, and employee capabilities. Four propositions are presented with authors arguing for integration of contextual factors in future research on complex service innovation processes.

Chapter 10

The tenth chapter, authored by Pikkarainen and colleagues titled "Needs and Implications of Data in Healthcare-Related Policymaking" explores the digitalization in the health sector, and offer insights regarding integration of data for developing preventive health care services. It examines the phenomena using a qualitative approach involving key municipal decision makers. The scholars suggest implications of data integration as a complex process but with an impactful outcome that could upgrade the level of health and wellbeing management beyond the current approaches.

References

Aas, T. H. (2016). Open service innovation: The case of tourism firms in Scandinavia. *Journal of Entrepreneurship, Management and Innovation*, 12(2), 53–76.

Aas, T. H., Breunig, K. J., & Hydle, K. M. (2017). Exploring new service portfolio management. *International Journal of Innovation Management*, 21(06), 1750044.

Aas, T. H., & Jørgensen, G. (2016). Business models for new experiential services. In: H.C.G. Johnsen, E. Hauge, M.L. Magnussen, and R. Ennals (eds.), *Applied Social Science Research in a Regional Knowledge System*. Routledge, New York, pp. 252–266.

Baines, T., Ziaee Bigdeli, A., Bustinza, O. F., Shi, V. G., Baldwin, J., & Ridgway, K. (2017). Servitization: Revisiting the state-of-the-art and research priorities. *International Journal of Operations & Production Management*, 37(2), 256–278.

Baregheh, A., Rowley, J., & Sambrook, S. (2009). Towards a multidisciplinary definition of innovation. *Management Decision*, 47(8), 1323–1339.

Barras, R. (1986). Towards a theory of innovation in services. *Research Policy*, 15(4), 161–173.

Bogers, M., Zobel, A. K., Afuah, A., Almirall, E., Brunswicker, S., Dahlander, L., ... & Hagedoorn, J. (2017). The open innovation research landscape: Established perspectives and emerging themes across different levels of analysis. *Industry and Innovation*, 24(1), 8–40.

Carlborg, P., Kindström, D., & Kowalkowski, C. (2014). The evolution of service innovation research: A critical review and synthesis. *The Service Industries Journal*, 34(5), 373–398.

Chesbrough, H. W. (2003). The era of open innovation. *MIT Sloan Management Review*, 44(3), 35–41.

Chesbrough, H. W. (2011). Bringing open innovation to services. *MIT Sloan Management Review*, 52(2), 85.

Chesbrough, H., Vanhaverbeke, W., & West, J. (2006). *Open Innovation: Researching a New Paradigm*. London, England: Oxford University Press.

Dahlander, L., & Gann, D. M. (2010). How open is innovation? *Research Policy*, 39(6), 699–709.

Den Hertog, P., Van der Aa, W., & De Jong, M. W. (2010). Capabilities for managing service innovation: Towards a conceptual framework. *Journal of Service Management*, 21(4), 490–514.

Drejer, I. (2004). Identifying innovation in surveys of services: A Schumpeterian perspective. *Research Policy*, 33(3), 551–562.

Dodgson, M., Gann, D., & Salter, A. (2006). The role of technology in the shift towards open innovation: The case of Procter & Gamble. *R&D Management*, 36, 333–346.

Dotzel, T., Shankar, V., & Berry, L.L. (2013). Service innovativeness and firm value. *Journal of Marketing Research*, 50(2), 259–276.

Edvardsson, B., & Olsson, J. (1996). Key concepts for new service development. *Service Industries Journal*, 16(2), 140–164.

Froehle, C. M., & Roth, A. V. (2007). A resource-process framework of new service development. *Production and Operations Management*, 16(2), 169–188.

Gallouj, F., & Djellal, F. (Eds.) (2010). Introduction: Filling the innovation gap in the service economy — A multidisciplinary perspective. In: *The Handbook of Innovation and Services: A Multidisciplinary Perspective*, Edward Elgar, Cheltenham, pp. 1–23.

Gallouj, F., & Windrum, P. (2009). Services and services innovation. *Journal of Evolutionary Economics*, 19(2), 141–148.

Gassmann, O. (2006). Opening up the innovation process: Towards an agenda. *R&D Management*, 36(3), 223–228.

Grönroos, C. (2007). *Service Management and Marketing: Customer Management in Service Competition*. John Wiley & Sons, Chichester.

Hipp, C., & Grupp, H. (2005). Innovation in the service sector: The demand for service-specific innovation measurement concepts and typologies. *Research Policy*, 34(4), 517–535.

Huizingh, E. K. (2011). Open innovation: State of the art and future perspectives. *Technovation*, 31(1), 2–9.

Hydle, K. M., Aas, T. H., & Breunig, K. J. (2014). Strategies for service innovation: Service innovation becomes strategy-making. In: A. Mention and M. Torkkeli (eds.), *Innovation in Financial Services: A Dual Ambiguity*. Cambridge Scholars Publishing, Newcastle, pp. 239–259.

Kuester, S., Schuhmacker, M. C., Gast, B., & Worgul, A. (2013). Sectoral heterogeneity in new service development: An exploratory study of service types and success factors. *Journal of Product Innovation Management*, 30(3), 533–544.

Lundvall, B. (1985). *Product Innovation and User-Producer Interaction*. Aalborg: Aalborg University Press.

Menor, L. J., Tatikonda, M. V., & Sampson, S. E. (2002). New service development: Areas for exploitation and exploration. *Journal of Operations Management*, 20(2), 135–157.

Merchant, K. A., & Van der Stede, W. A. (2013). *Management Control Systems: Performance Measurement, Evaluation and Incentives*. Pearson Education.

Mina, A., Bascavusoglu-Moreau, E., & Hughes, A. (2014). Open service innovation and the firm's search for external knowledge. *Research Policy*, 43(5), 853–866.

Perks, H., & Riihela, N. (2004). An exploration of inter-functional integration in the new service development process. *Service Industries Journal*, 24(6), 37–63.

Pittaway, L., Robertson, M., Munir, K., Denyer, D., & Neely, A. (2004). Networking and innovation: A systematic review of the evidence. *International Journal of Management Reviews*, 5(3–4), 137–168.

Powell, W. W., Koput, K. W., & Smith-Doerr, L. (1996). Interorganizational collaboration and the locus of innovation: Networks of learning in biotechnology. *Administrative Science Quarterly*, 41(1), 116–145.

Randhawa, K., Wilden, R., & Hohberger, J. (2016). A bibliometric review of open innovation: Setting a research agenda. *Journal of Product Innovation Management*, 33(6), 750–772.

Santonen, T., & Conn, S. (2015a). Research topics at ISPIM: Popularity-based scientometrics keyword analysis. In Huizingh, Eelko, Torkkeli, Marko, Conn, Steffen and Bitran, Iain (eds.), *The Proceedings of The XXVI ISPIM Innovation Conference*, 14–17 June, Budapest, Hungary.

Snyder, H., Witell, L., Gustafsson, A., Fombelle, P., & Kristensson, P. (2016). Identifying categories of service innovation: A review and synthesis of the literature. *Journal of Business Research*, 69(7), 2401–2408.

Sundbo, J. (1997). Management of innovation in services. *Service Industries Journal*, 17(3), 432–455.

Van Ark, B., Broersma, L., & den Hertog, P. (2003). Service innovation, performance and policy: A review. *Synthesis Report in the Framework of the Project Structural Informatievoorziening in Diensten* (*SIID*). Ministry of Economic Affairs, The Hague.

West, J., & Bogers, M. (2014). Leveraging external sources of innovation: A review of research on open innovation. *Journal of Product Innovation Management*, 31(4), 814–831.

https://doi.org/10.1142/9789811234491_0002

Chapter 2

Open Service Innovation: A Systematic Literature Review

Tor Helge Aas[*,†,§], *Anne-Laure Mention*[‡,¶] *and Christina Lee*[‡,||]

[*]*School of Business and Law, University of Agder, Campus Kristiansand, Universitetsveien 25, 4630 Kristiansand, Norway*

[†]*NORCE Norwegian Research Centre, Nygårdsgaten 112, 5008 Bergen, Norway*

[‡]*RMIT University, School of Management, Melbourne, Australia*

[§]*tor.h.aas@uia.no*

[¶]*Anne-laure.mention@rmit.edu.au*

[||]*theartofchristinalee@gmail.com*

Abstract. Over the last years, the phenomenon of open service innovation has gained an increasing amount of attention from researchers and practitioners. In this chapter, we take stock of the emerging literature on open service innovation and provide a systematic review of this literature. We identify and analyze 171 peer-reviewed articles published in 65 journals between 2003 and 2018. The analysis shows that the current literature has discussed topics such as co-creation, collaboration, absorptive capacity, crowdsourcing and citizen participation in relation to open service innovation. However, specific theoretical development related to open service innovation is scarce, and this constitutes a gap in the literature. We argue that continued empirical exploration of open service innovation is a viable research area and suggest that the development of an open-dominant logic framework, which follows along the same vein as service-dominant logic, could be valuable as a theoretical underpinning of future research.

Keywords. Open service innovation; systematic literature review; co-creation, collaboration for service innovation; citizen participation; crowdsourcing; absorptive capacity.

1. Introduction

Open service innovation has captured interest of academics, managers and policymakers alike as economies continue to shift the focus from product-based innovation to service oriented innovations (Love *et al.*, 2011). Innovation in the service sector has been prominent in not only western service-based economies like Australia but also eastern traditionally manufacturing-based economies such as China (Chesbrough, 2011). However, the problem is that "we know much less about how to innovate in services than about how to develop new products and technologies" (Chesbrough, 2011, p 2).

The increasing interest in open service innovation has been stimulated by several factors including (1) shifting economic trends as services now comprise of more than 70% of economic activity (OECD, 2000), (2) firms sourcing knowledge outside their organizational boundaries to remain competitive or gain competitive advantage (Chatfield & Reddick, 2017; Asikainen & Mangiarotti, 2017; Loukis *et al.*, 2017; Martovoy *et al.*, 2015), and (3) research interest towards innovations in service firms (Cardellino & Finch, 2006; Djellal & Gallouj, 2001; Drejer, 2004; Evangelista, 2000; Miles, 2000; Sundbo, 1997; Tether, 2005). Consequently, the literature on — innovation in services has emerged from "a neglected and marginal status to achieving wide recognition as being worthy of depth of study" (Miles, 2000). Perhaps the challenge is that much of the open innovation literature has origins in innovation theory which has roots in a time when manufacturing was still a major economic activity (Gadrey *et al.*, 1995). In most cases, the studies conducted are of private manufacturing firms or they are so diversely covered in academic journals that it remains unclear how open service innovation should be defined, developed and managed.

A critical reading of the literature highlights that open service innovation is often conceptualized with open innovation, that is to "employ purposive inflows and outflows of awareness to step up internal innovation and develop the markets for external use of innovation" (Foroughi *et al.*, 2015, p. 692). Despite this, studies that relate to open innovation within services have largely relied on cases in Information and Communication Technology (ICT) sector, emphasizing the service aspects in manufacturing sector and not on services within service firms. More recent scholarship in the area has explored the implications of open innovation paradigm in the context of services (see Chesbrough, 2011; Martovoy & Mention, 2013; Martovoy *et al.*, 2015). Yet, there is still a disparity in the sample cases — i.e., they are mainly from product-based companies which are aggregating services to their portfolio (Aas & Pedersen, 2016). Service innovation differs to manufacturing in the open arena. For instance, service firms tend to focus on customers as the main source of external knowledge when seeking to achieve service innovations (Mina *et al.*, 2014).

The goal of this chapter thus, is to specify the different (multifaceted) dimensions of open innovation in services, identify research gaps in the literature and provide avenues for further research. As such, the following research question shapes the remainder of this chapter: "What are the multifaceted approaches to open innovation in services and how does open innovation impact the development of new services?" This specific focus was chosen because understanding open service innovation and how to implement open service innovation processes is often seen as a way for firms to escape the commodity trap and reach levels of success they never expected (Chesbrough, 2011, pp. 2–6). Moreover, through greater understanding of the different approaches to open service innovation, firms might be able to improve their competences, capabilities and resources, which in turn could enable them to develop business models that strengthen their capacity to innovate. We address the research question by conducting a systematic literature review. Given open service innovation is part of a widely diverse field in need of succinct exploration, a systematic review was conducted to develop an understanding of the current research being conducting in open service innovation. A systematic review helps to form the basis for developing guidelines and identify any gaps in knowledge about the specific field of study (Shamseer *et al.*, 2015). It also synthesizes research according to explicit and reproducible methodology (Cassell *et al.*, 2006), removing bias and consolidating existing research through a rigorous and comprehensive review process.

This chapter contains the following sections. First, we draw on current theory to help form definitions of open innovation and service innovation that will be used in this chapter. Second, the research method for the systematic literature review adopted by this chapter is detailed followed by a discussion of the findings and how they relate to the existing body of knowledge. Finally, the chapter concludes with a summary of managerial implications, and avenues for future research.

2. The Concept of Open Innovation

A significant body of literature exists describing the characteristics of open innovation and how it is used as a way for firms to gain competitive advantage. Chesbrough (2003), an academic considered as the forefather of open innovation, first introduced this concept in his seminal book titled *Open Innovation*. Chesbrough (2003) describes open innovation as a paradigm that can be understood as the antithesis to internal research and development activities. He further explains that open innovation assumes that firms can and should use external and internal knowledge and ideas to advance their technology. Chesbrough also notes that open innovation explicitly incorporates the business model as the source of both value creation and value capture (Chesbrough, 2011). To simplify his explanation of open innovation, Chesbrough provides a succinct definition that has since informed research in the area: "open innovation is the use of purposive inflows and outflows of knowledge to accelerate internal innovation, and expand the markets for external use of innovation, respectively" (Chesbrough, 2011, p. 2).

The idea of open innovation was therefore "to take the traditional innovation process and open it up at all stages as the process of innovation flows from the laboratory to market" (Chesbrough, 2017, p. 29). This flow involved three key types of process: The first being *outside–in* where firms can tap into external knowledge sources such as customers, suppliers, competitors, and public and commercial research institutions (Aitamurto & Lewis, 2012; Enkel *et al.*, 2009) to increase their innovative outcomes. Firms that use the outside–in process become less dependent on their internal R&D as they leverage off other discoveries to organizations they are engaged, or sometimes less engaged, with. The outside–in process has also fuelled an increasing awareness of the importance of innovation networks (Dittrich & Duysters, 2007; Chesbrough & Prencipe, 2008), new forms of customer integration, such as crowdsourcing (Davis *et al.*, 2015), mass customization, and customer community integration (Piller & Fredberg, 2009), alongside highlighting the role of innovation intermediaries (Aquilani *et al.*, 2016; Dong & Pourmohamadi, 2014).

The second is *inside–out*, a less utilized process, which is when firms focus on externalizing the innovation process by sharing knowledge in order to bring ideas to the market at a faster rate (Enkel *et al.*, 2009). Chesbrough (2017, p. 29) explains that not all firms' innovations make it to market, in fact some are abandoned due to cost, resource or because they have been explored to in the full extent of firm capability but have not reached a conclusive innovation — these are referred to as "false-negative projects". When a firm releases these projects and related intellectually property (IP) to the market for others to innovate with, some of them can turn into enormously valuable endeavours. For instance, Amazon shared its knowledge in e-commerce and helped third-party retailers develop their own merchandising sites — consequently Amazon began hosting these retailers on its servers, adapting their business model to now account for being paid for sharing its knowledge in the market (Aitamurto & Lewis, 2012).

The third type is a *coupled process* where a firm creates alliances, partnerships and joint ventures to create value for both parties involved (Aitamurto & Lewis, 2012; Enkel *et al.*, 2009). The coupled process refers to co-creation as firms combine both the outside–in and inside–out processes to jointly develop and commercialize innovation (Enkel *et al.*, 2009). Co-creation is the interplay between internal and external actors during the process of opening up firm's activities to other entities such as customers and other stakeholders (Ramaswamy & Ozcan, 2018). Compared to the other two processes, the coupled process is under-examined in extant literature on open innovation but remains promising (Aitamurto & Lewis, 2012).

As researchers continue to explore open innovation, the definition of the concept continuously evolves to embrace new concepts and theories. Chesbrough recognized this and noted that "we now define open innovation as a distributed innovation process based on purposively managed knowledge flows across organizational boundaries, using pecuniary and non-pecuniary mechanisms in line with the organization's business model" (Chesbrough, 2017, p. 30). In other words, firms seeking to collaborate with others, be it with internal or external individuals, organizations or institutes, can create and capture value through open and collaborative knowledge exchange mechanisms in the process of innovation and co-creation (Aitamurto & Lewis, 2012; Mention, 2011). There are many ways for a firm to open up their innovation processes, however the main idea is that firms utilize inside–out and outside–in practices in order to open up their innovation process (Gianiodis *et al.*, 2014; Grøtnes, 2009; Van de Vrande *et al.*, 2009).

3. The Concept of Service Innovation

Researchers across many fields, such as marketing (Berry *et al.*, 2006; Oliveira & Von Hippel, 2011), economics (Cainelli *et al.*, 2006; Gallouj, 2002; Gallouj & Savona, 2009), information systems (Alter, 2008; Lyytinen & Rose, 2003; Rai & Sambamurthy, 2006), operations (Edvardsson & Olsson, 1996; Fitzsimmons & Fitzsimmons, 2000; Metters & Marucheck, 2007; Oke, 2007), and strategy (Dörner *et al.*, 2011), have highlighted the significance of service innovation. However, theory building around the concept is still novel with vague and dispersed definitions of the core concepts (Carlborg *et al.*, 2014; Witell *et al.*, 2016). In response to the diversity found in service innovation research, Coombs and Miles (2000) developed three distinct perspectives, assimilation, demarcation, and synthesis which claim to be separate and distinct, categorizing existing research into one of these perspectives (Witell *et al.*, 2016). These perspectives are used to help formulate a clear understanding of the service innovation research and provide insight into the theoretical concepts of service innovation. Thus, assimilation, demarcation and synthesis are described here to help provide a clearer understanding of service innovation.

Studies categorized into the assimilation perspective are numerous and focus on the impact of new technology (Witell *et al.*, 2016). The core construct of assimilation is that service innovation is fundamentally like manufacturing innovation (Coombs & Miles, 2000) and adapts the same theories and instruments used for traditional product innovation research with no modification or adjustment (Gallouj & Savona, 2010; Witell *et al.*, 2016). This traditional approach to innovation in services only considers technological or "visible" modes of innovation and ignores other non-technological or "invisible" modes of innovation (Morrar, 2014). Therefore, other innovation in service activities such as, social innovation, organizational innovations, methodological innovations, marketing innovations and innovations involving intangible products of processes are underestimated (Morrar, 2014). Another important supposition of assimilation is that the service sector is becoming more technology and capital intensive (Gallouj & Savona, 2009).

In contrast to assimilation, the demarcation perspective argues that service innovation is highly distinctive and requires novel theories and instruments. The demarcation perspective suggests that service innovation is fundamentally different in all aspects from product innovation (Coombs & Miles, 2000; Morrar, 2014; Witell *et al.*, 2016), with scholars arguing that innovation studies have failed to recognize the specificities of services and the important contributions that services make to products (Witell *et al.*, 2016). Furthermore, research within this perspective challenges the theoretical foundation for innovation studies and illuminates important elements that were previously neglected (Witell *et al.*, 2016). For instance, Drejer (2004) argued that one of the important factors of the demarcation perspective is that it expanded on what can be considered an innovation. It focuses on non-technological (service-based) and invisible innovation output and processes, such as the intangible nature of services, customer integration, and new organizational structures (Morrar, 2014; Witell *et al.*, 2016). Although the demarcation perspective is not fully developed, it has been the underpinning of specialized studies of innovation in services (Coombs & Miles, 2000).

Finally, the synthesis or integrative perspective aggregates and critiques both the assimilation and demarcation perspectives with the idea to provide a broader view of service innovation which encompasses both services and manufacturing (Morrar, 2014; Witell *et al.*, 2016). This perspective considers both services and goods, including technological and non-technological modes of innovation (Gallouj & Savona, 2009; Morrar, 2014) and represents the emerging theoretical development in service innovation discussion (Morrar, 2014). Importantly, the synthesis perspective is motivated by the convergence between service and manufacturing, that is, it identifies that two main changes taking place: "manufacturing is becoming more like services and services are becoming more like manufacturing" (Morrar, 2014, p. 10). This on the one hand is due to manufacturing firms producing service products and an increasing firms' turnover through selling services (Howells, 2006). On the other hand, service firms are becoming more innovative and therefore greater parts of their innovation output are reflected by traditional technological innovation in manufacturing (Morrar, 2014). This synthesis perspective highlights the multifaceted and complex operating conditions of modern business, be it service-oriented or service-focused (Salter & Tether, 2006). In sum, service innovation is still a growing field of research and although these three perspectives play an active role in helping to decipher the theoretical framework for service innovation, further research into this area can offer additional insight into the concept.

4. Methodology

A systematic literature review aims to locate, select and synthesize themes of a specific literature topic with a purpose of providing a theoretical and substantial contribution to the analyzed field (Marabelli & Newell, 2014). This chapter follows the principles described by Cassell *et al.* (2006) for producing a transparent, inclusive, explanatory and heuristic systematic review.

This literature review identifies publications of interest to the researchers and then describes the process in which the research includes and excludes relevant publications in preparing for the research report (Savino *et al.*, 2017). The search strategy focused on peer-reviewed papers identified through an electronic search of the Scopus database. To set review boundaries, the search of peer-reviewed articles was limited to articles published between 2003 and 2018. This captures a time of growing interest and popularity for research and theory development in open innovation (Carroll & Helfert, 2015; Chesbrough, 2017) since it was first coined by Chesbrough (2003). To locate terms used by researchers to explore open innovation in services, a basic search string of ("open innovation in services" OR "open service innovation") was utilized. The results of this search string were too narrow and lacked depth; producing 20 articles, 14 book chapters, three conference papers, one book and one article in press. Thus, a secondary search was conducted with refined keywords (open + service* (e.g., services) + innovation) which produced 2,871 document results. As part of the exclusion criteria, only peer-reviewed articles were used in this research leaving 1,278 document results for title and abstract analysis.

Following the requirement for transparency in a systematic review, a thorough evaluation of article titles and abstracts was undertaken by subjecting each article to a series of explicit selection criteria. The selection criteria aimed to exclude studies that had minimal reference to open innovation or open service innovation and included studies that explicitly referenced innovation in an open context or innovation in services. This was achieved by using the word "filter" tool within Scopus which highlighted how many times the search terms appeared in the title or abstract. This resulted in a smaller sample of articles, largely due to being unrelated to open service innovation (e.g., they were solely about services or referenced key terms only once or twice). However, noting that some articles referred to both product and service innovation, exceptions to the inclusion and exclusion criteria were developed. In some cases, abstracts made it difficult to identify the aims, approach and findings of the study therefore introductions and/or conclusions were examined in those cases. After removing duplicates, 171 articles were left for further analysis. This analysis included a full article screening, checking for relevance to the research topic and sorting articles into category groups (open innovation; co-creation; collaboration; service innovation; etc.). Once the review was completed 87 articles were left and formed the selected sample for this review. The 87 articles were then organized according to the classification variables, for example: "Year of Publication", "Research method", "Country/ies of study"; "Research sector"; "Research Journal"; "Qualitative/Quantitative/Conceptual"; "Main theories"; "Research question and Hypotheses"; "Methods"; and "Author location". The extracted data, and the main findings of the studies, were then carefully recorded into Microsoft Excel for further analysis and data visualization.

5. Findings

5.1. *Overview of the literature*

In a bibliometric review conducted by Randhawa *et al.* (2016), it was identified that the open innovation field attracted considerable attention from academics following Chesbrough's (2003) seminal work on the topic. Moving forward, researchers begun to develop theories and ideas around open innovation in services, seeking to move away from product innovation and into a service focused concept. Although a steady shift our findings indicate that researchers started to pay attention to open innovation in services mainly from 2009, with a peak number of articles published in 2013 and 2014 (n = 28) which can perhaps be related to Chesbrough's work in 2011, sparking a greater interest in the field, however there has been a slight decline since 2014. It is possible that given open service innovation is an emerging field of study, some of the literature is not captured in this review due to being published in non-peer-reviewed outputs or published in other forms such as grey literature. It is likely that new theories will be developed as new research is conducted which can either lead to new revelations into the open service innovation process or further distort the field of study.

The period of 2003 and 2018 account for most of the research outputs, many of which are empirical studies (n = 68) published in innovation and/or management journals, with few found in non-innovation-centric journals. Randhawa *et al.* (2016) explain that this is of concern as it indicates that open innovation research is a closed affair and limited from influencing external research fields. They further argue that research into open innovation is inward focus, however the essence of open service innovation suggests cross-disciplinary work to expand knowledge outside the historical view of innovation that was mostly centred on manufacturing firms.

Notably, most of the literature on open service innovation concentrates on the private sector as empirical setting and authors are predominately based in the United States (n = 12, 12.37%), Germany (n = 9, 9.28%), Italy and Sweden (n = 8, 8.25%). A little over 19% (n = 17) are related to the public sector, compared with just over 37% (n = 33) related to the private sector (the remainder are either both or unspecified). Further, the study concluded with 69 separate journals of which 59 journals provided only one article each and the remainder 10 journals providing two or more articles. Key publications were found in the *International Journal of Technological Innovation, Entrepreneurship and Technology Management* (three articles), *Journal of Technology Management & Innovation* (three articles), *Journal of Product Innovation Management* (three articles), and the *European Journal of Innovation Management* (three articles) with 16.09% published in 3/4* ranked journals, 17.24% in 2 ranked journals, 9.20% in 1 ranked journals and 35.63% were unranked (accordance with the Association of Business Schools journal ranking). The study sample included 18 conceptual studies, 26 quantitative studies, three mixed method studies and 39 qualitative studies. Thus, this review illustrates that open innovation in services is being explored conceptually, quantitatively and qualitatively, and the topic has international appeal with studies conducted in several countries. It also highlights that the research is still in an exploratory stage with many of the articles being inductive in nature.

Interestingly, observations related to the 45 unspecified articles suggest that, researchers have chosen to explore open innovation and open innovation in services in a broader sense. Many of the unspecified articles do not specifically refer to one organization as their subject of analysis, rather researchers look at open innovation in services or open service innovation as a general concept. Of the articles that look at industry specific open innovation, they focus on how an organization within the industry (i.e., public sector, finance, tourism) can benefit from applying openness.

Although there are several industries researched by academics, there is a clear focus on the information technology (IT) industry (n = 19) along with many papers using case studies where firms have implemented software in order to achieve beneficial outcomes. Although open source software makes up part of the research conducted within the IT industry, there was also a clear interest in the innovation of products and services within IT firms and how open innovation interplays with product and service innovation. Perhaps, it is worth noting that in practice open service innovation is developing across multiple sectors and is no longer remaining in the realm of the private sector. Specifically, governments now use open service innovation techniques to develop positive citizen engagement and inclusive policies for community developments. This kind of citizen innovation is best described as crowdsourcing and involves large numbers of citizens working together with government to produce a socially accepted result.

Furthermore, the analysis revealed that researchers came from a diverse range of countries, but they tend to predominately conduct studies in western countries or in a general setting that is unspecific to location. While the former adds to the concerns for under representation of less developed economies, the latter is indicative of the changes occurring in the economy which is pushing for new ways to compete on a global scale, highlighting that open service innovation might be connected to the big ideas in business today (e.g., globalization and global citizenship).

5.2. *Content analysis*

Our findings indicate that studies typically look at different components of open innovation and study their effectiveness in service firms rather than discussing open service innovation in a broad sense. Examples of such components are: (1) crowdsourcing ideas for service and product innovation; (2) citizen innovation in the public sector, (3) government and major cities; (4) open-source innovation for use in online companies or to connect firms to other likeminded firms to innovate with; (5) innovation intermediaries which match stakeholders to a firm wanting to innovate outside their organizational boundaries; (6) co-creation with customers to leverage of the knowledge they have; (7) collaboration between stakeholders — deciphering who are the best stakeholders to engage in; (8) innovation networks where likeminded people can openly innovate together; (9) openness and the degree in which it effects innovation performance; and (10) finally when to be open vs. closed in innovation activities. This has resulted in studies into open service innovation being widely diverse and sometimes neglected in prior literature reviews due to a limitation in research search scope. Further analysis of the content of the reviewed articles, however, revealed that there are five main areas of study that academics examining open service innovation refer too: The first is co-creation and collaboration, the second is absorptive capacity, the third is crowdsourcing and citizen participation in public-sector innovation, the fourth is innovation performance, and the fifth area of discourse has been centred on product vs. service innovation and the changes that can or should occur when implementing open service innovation.

In the first area, co-creation and collaboration is often examined through the lens of SD-logic, which can be summarized as the exchange of knowledge and application of specialized skills where service is exchanged (Lusch & Vargo, 2006). Put simply, SD logic is a reflection of the transition from the industrial era to a service-driven era (Lusch & Vargo, 2006), thus it is apparent why SD logic forms part of the literature around open service innovation. Mele and Russo-Spena (2017) discuss that the conceptualization of innovation in terms of SD logic is primarily based on the different meanings of service. They explain that unlike goods-dominant logic, SD logic focuses on service as a process not a unit of output and Lusch *et al.* (2008) add that SD logic focuses on dynamic resources such as knowledge and skills rather than static resources along with the collaborative process between the firm and its customers rather than that is manufactured and supplied. Where open service innovation and SD logic intersect greatly is in the co-creation between firm and customer. Both define a process wherein the firm and customer work together to seek ways to create value; as such the customer is seen as a co-creator of value (Lusch & Vargo, 2006). Filieri (2013) further emphasizes that value is co-created by firms, employees, customers, stockholders, and government agencies, but is always determined by the beneficiary which is often the customer. In the literature which discussed co-creation, firms are always co-creating with their customer, and SD logic theory is largely drawn upon when explaining the customer value creation. This does not come as a surprise, aside from firms co-creating with their customer to create new service developments, SD logic stresses that successful innovation relies on all participants collaboratively co-creating value. In terms of open service innovation, this can refer to many stakeholders such as universities, competitors, supplies and internal knowledge supporters — the reason for this collaboration and co-creation is that firms rarely possess enough knowledge and sufficient resources to create the innovation that is needed for them to compete globally (Mele & Russo-Spena, 2017). Further, SD logic strongly emphasizes the relationship between knowledge and skills and the interaction between mutually beneficial relationships that improve adaptability and survivability for those involved (Vargo *et al.*, 2008), for example those involved in open innovation activities with others.

In terms of collaboration, if was revealed that several studies discussed how collaboration among internal and/or external stakeholders has an effect on innovation performance and at what rate success is achieved. However, researchers have argued that collaboration is an understudied area of open innovation which has resulted in a lack of understanding in a practical sense as well as theoretical one which can often lead to cost intensive and unsuccessful new service development (Temel *et al.*, 2013). What this literature review can confirm however is that there is evidence to suggest that collaboration has an impact on innovation performance and that impact differs with each innovation project (Temel *et al.*, 2013). For example, Levine and Prietula (2014) explain that there are three elements that affect performance where collaboration is concerned within open innovation: the cooperativeness of participants, the diversity of their needs, and the degree to which the goods are rival. As much of the success relies on the effectiveness of the collaboration between participants, their knowledge and skills contributed, and the ability to create something novel, we begin to discover a reoccurring theme within the literature; knowing how individuals interact when creating knowledge collaboratively (Du Chatenier *et al.*, 2009). This leads into further expansion of the open innovation area which look at open innovation team dynamics and success along with how firms need to support their internal innovators to be able to contribute effectively to the innovation process. This human side of open innovation needs further research (Bogers *et al.*, 2017).

The second area identified by this review was related to absorptive capacity. This study found that absorptive capacity, which is the ability of a firm to identify, assimilate, transform, and apply valuable external knowledge into its practices, has begun to surface as an area of open service innovation required for successful innovation to occur. What was identified here is that a small number of studies examine absorptive capacity and discuss the firm's ability to absorb the knowledge they receive — they stress the importance of developing the firm's capability and managers' characteristics and practices (Pedrosa *et al.*, 2013). Often it is expressed that there is a need for firms to cultivate open innovation business models which promote a greater appreciation of external knowledge transfer, this includes assessing the managers' characteristics and practices that capture the dominate pattern of processing external knowledge in open innovation. However, data collected within this research suggests that the literature on absorptive capacity and collaboration fails to cope with the emerging requirements of innovation projects and thus more research into how collaboration effects open service innovation and how absorptive capacity effect innovation outcomes in firms is needed. Furthermore, as open service innovation is predominately characterized by individuals and teams that provide knowledge and resources there lacks clear derivatives of the human side to open innovation. This can have implications for policy and decision-makers as the way in which open innovation is managed could mean the success or failure of an innovation project.

The third area that had a reoccurring theme within the open service innovation literature was crowdsourcing and citizen participation in innovation whereby, mostly government, cities and other public-sector organizations, invited citizens to participate in ideation and the co-creation for new service development and improvements. Numerous papers cited the term crowdsourcing and citizen participation as a key element of open service innovation (Mergel & Desouza, 2013; Davis *et al.*, 2015). Although there is a direct connection between crowdsourcing and the private sector with many private firms using crowdsourcing to innovation, the emergence of citizen and user participation in innovation activities indicates that the landscape is changing from private manufacturing and service firms to include public organizations. Perhaps the gap between public and private sector is starting to level out. Within this finding, an assessment of the disparity between private-sector innovation characteristics and public-sector innovation indicates that the same principles that are seen in the private sector may not directly and successfully translate into the public domain, a notion that needs further case-based exploration.

The fourth area that academics consider is innovation performance which captures the relationship between new product or service innovation and firm outcomes (Cheng & Huizingh, 2014). A large body of literature confirms that open innovation practices are beneficial for business growth and performance and therefore open innovation should be considered within a firm's strategy. What is unclear is what kind of organizational context suits open innovation best and what open innovation activity; inside–out, outside–in and coupled activities are best for the firm seeking to innovate (Cheng & Huizingh, 2014). Further to this, academics have looked at multiple levels of open innovation in assessing performance, for example the effect of competitive intensity, involvement in innovation activities, participation in open networks, internal R&D activities and external knowledge sourcing. Very few of these academics have concluded that open service innovation does not positively effect innovation performance, however further understanding of the process in which service firm's innovation needs to be discussed.

The fifth area of discourse was centred on product vs. service innovation. Rather than focusing on traditional service firms, researchers in this area discussed how services can help firms that mainly developed products to become more competitive through the implementation of services that complement their product offering. Where the difference lies is in the nuances of service — looking back to SD logic theory, it best explains the term service and services which can be used here. The use of "service" means "a process of doing something for someone", rather than the plural "services" which is implying units of output which is a goods-dominant logic term (Lusch & Vargo, 2006). It is important to understand that service firms require different approaches to open innovation than their manufacturing counterpart, for one they rarely have any product offerings or if they do it is not as part of their core framework.

Although service firms can engage external parties in the same way as manufacturing firms, the level of engagement with these parties varies greatly and with whether a firm chooses inside–out or outside–in approach. Where manufacturing firms can invite suppliers, competitors, and consultants (for example) to help find innovative solutions, service firms predominately call upon their customer as their main contributor to the innovation process across its various stages (Martovoy & Mention, 2016). For service firms, the purpose of open service innovation is to improve customer satisfaction by providing a service that is unique to their wants or needs. By involving the customer in the innovation process, service firms can foster greater engagement with their offerings and subsequently, loyalty to their brand and thus competitive advantage.

Where the literature lacks, is in explaining this difference and examining open service innovation more closely from a service firm perspective. There are some studies that look at this in more depth and they help to build on the foundations of open service innovation, such as Martovoy and Mention (2014, 2016) who discuss open innovation within financial services. They highlight that empirical studies suggest financial services tend to rely on inflows of knowledge and that this knowledge originates from different external stakeholders, for example, financial services can benefit and take advantage of their competitors or third-party providers of other services. However, Martovoy and Mention (2016) suggest that consultants, suppliers and universities have yet to feature within the literature for open service innovation within financial services. This corresponds with this literature review, little is known about service firms and who they collaborate with, outside of the customer, and at what stages of the innovation process.

6. Discussion

Open innovation was first introduced over a decade ago by Chesbrough (2003) and has since seen exponential growth in the academic field. Many researchers have developed theories, hypothesis, concepts and practices to help explain open innovation, but this rapid growth created confusion and distortion within the field (Chesbrough, 2017). Adding further complexities to the field is the introduction of open service innovation which specifically addresses service firm innovation, highlighting that service firms are increasingly innovating with stakeholders outside of their organizational boundaries, much like their manufacturing counterparts. However according to Chesbrough (2017) much of the literature seems to substitute a piece of the field for the whole. Thus, even though open service innovation is gaining increased interest, much of the developments around open service innovation are skewed to traditional theories and methods relating to product innovation, innovation in manufacturing firms, or R&D intensive innovation.

This systematic literature review has identified that further research into open service innovation is needed. It revealed the multiple facets of open innovation used in conjunction with open service innovation and that scholars are yet to identify theoretical foundations that can explain open service innovation. Arguably, this could be due to the infancy of open service innovation research as much of the current literature is conceptual and explorative. However, to help further the research being conducted within open service innovation, it is plausible that new theories need to be developed which specifically address the key components of open innovation. Developing a theory akin SD logic, for example, could help to establish a solid foundation for open innovation research along with helping to consolidate the research being undertaken in the field. The choice of SD logic to help develop a new theory is twofold. Firstly, SD logic resonates with open service innovation at a fundamental level. It works toward stepping away from goods-dominate logic and pushes for service in exchange for service, a concept like what is found in open service innovation. Secondly, SD logic explores concepts that align with open innovation practices, for example, value co-creation and the act of seeking information from the market rather than solely within the firm's walls. Given SD logic refers much to marketing rather than business models and firm behaviors around innovation and competitive advantage; it is proposed that the new theory in question should be firm centric, open focused and dynamic. We propose an open dominate theory which addresses how firms can open their innovation processes in effective and manageable ways. This theory should also address when a firm should be open and when to be closed as this is in important key element that still needs to be addressed in further research if open service innovation is to develop into a well-rounded concept. If developed holistically, taking into consideration other theoretical concepts identified in this research, an open dominant theory could potentially become the theoretical foundation for further research into open service innovation and equally could bring consistency and clarity across the field.

7. Conclusions

This chapter offers a systematic review of studies on open service innovation published from 2003 and 2018 in 65 journals. It analyses 171 articles and provides a clear picture of current trends and growth in the open service innovation literature suggesting some progress towards distinguishing the different characteristics of open service innovation. The review illustrates the diversity of open service innovation, identifying the need to consolidate and provide solid theoretical foundations specially for open service innovation. What the analysis shows is that current literature has focused and incorporated co-creation, collaboration and open innovation into open service innovation, but specific theoretical development of open service innovation is scares and this constitutes a gap in the research. This enforces the conclusion that there is significant scope to build upon existing theories to develop a theory that is uniquely applicable to the open service innovation context. It highlights that continued exploration of open service innovation is a viable research area and necessary in deepening our understanding of the concept.

Our attempt to fill the above gap involved the introduction of an open dominate logic which follows along the same vein as SD logic (Vargo & Lusch, 2017). The purpose of this new theory is to draw upon other theories such as, openness, co-creation as SD logic, to take the most relevant components and developing a theoretical foundation for open service innovation that will underpin future research and advancements in the field. This clearly identifies new avenues of research. First, a review of the main theories found in open service innovation need to be further unpacked and analyzed with an assessment made on what impact each theory has in open service innovation. Understanding which theories positively impact open service innovation and which do not provide much insight requires attention and deserves further research. Second, a meaningful avenue for further research might involve developing an understanding of open innovation in services across knowledge intensive versus non-knowledge intensive service sectors. This would promote more research that focuses on open innovation in service firms and sheds light on any process differentiation between service sectors.

This chapter also concluded that open service innovation is still in its conceptual stages of developments and much of the literature found is exploratory in nature. It is acknowledged that due to the emerging nature of open service innovation, there could be a larger body of research that isn't identified within the scope of this review, such as those found in grey literature, conference papers and published books. Thus, it should be considered that a further review, incorporating the above, should be conducted to expand and compliment the research provided here. The above avenues for further research suggest that, to move away from traditional theories and interpretations of open service innovation, scholars should look to examine new theories such as open dominate logic to develop further understanding of open service innovation. In turn, this would help in separating open innovation in services (which can include manufacturing firms) to provide a more structured and relevant body of research into how service firms innovation in an open way.

References

Aas, T. H., & Pedersen, P. E. (2016). The feasibility of open service innovation. In *Open Innovation: A Multifaceted Perspective: Part II*, pp. 287–314.

Aitamurto, T., & Lewis, S. C. (2013). Open innovation in digital journalism: Examining the impact of Open APIs at four news organizations. *New Media & Society*, 15(2), 314–331.

Alter, S. (2008). Service system fundamentals: Work system, value chain, and life cycle. *IBM Systems Journal*, 47(1), 71–85.

Aquilani, B., Abbate, T., & Dominici, G. (2016). Choosing open innovation intermediaries through their web-based platforms. *The International Journal of Digital Accounting Research*, 16, 35–60.

Asikainen, A. L., & Mangiarotti, G. (2017). Open innovation and growth in IT sector. *Service Business*, 11(1), 45–68.

Berry, L. L., & Shankar, V. & Parish, T. (2006). Creating new markets through service innovation. *Sloan Management Review*, 47(2), 56–63.

Bogers, M., Zobel, A. K., Afuah, A., Almirall, E., Brunswicker, S., Dahlander, L., Frederiksen, L., Gawer, A., Gruber, M., Haefliger, S., & Hagedoorn, J. (2017). The open innovation research landscape: Established perspectives and emerging themes across different levels of analysis. *Industry and Innovation*, 24(1), 8–40.

Cainelli, G., Evangelista, R., & Savona, M. (2005). Innovation and economic performance in services: A firm-level analysis. *Cambridge Journal of Economics*, 30(3), 435–458.

Carlborg, P., Kindström, D., & Kowalkowski, C. (2014). The evolution of service innovation research: A critical review and synthesis. *The Service Industries Journal*, 34(5), 373–398.

Cardellino, P., & Finch, E. (2006). Evidence of systematic approaches to innovation in facilities management. *Journal of Facilities Management*, 4(3), 150–166.

Carroll, N., & Helfert, M. (2015). Service capabilities within open innovation: Revisiting the applicability of capability maturity models. *Journal of Enterprise Information Management*, 28(2), 275–303.

Cassell, C., Denyer, D., & Tranfield, D. (2006). Using qualitative research synthesis to build an actionable knowledge base. *Management Decision*, 44(2), 213–227.

Chatfield, A. T., & Reddick, C. G. (2017). A longitudinal cross-sector analysis of open data portal service capability: The case of Australian local governments. *Government Information Quarterly*, 34(2), 231–243.

Cheng, C. C., & Huizingh, E. K. (2014). When is open innovation beneficial? The role of strategic orientation. *Journal of Product Innovation Management*, 31(6), 1235–1253.

Chesbrough, H. W. (2003). *Open Innovation: The New Imperative for Creating and Profiting from Technology*. Boston, Massachusetts: Harvard Business Press.

Chesbrough, H. W. (2011). Bringing open innovation to services. *MIT Sloan Management Review*, 52(2), 85.

Chesbrough, H. (2017). The future of open innovation: The future of open innovation is more extensive, more collaborative, and more engaged with a wider variety of participants. *Research-Technology Management*, 60(1), 35–38.

Chesbrough, H., & Prencipe, A. (2008). Networks of innovation and modularity: A dynamic perspective. *International Journal of Technology Management*, 42(4), 414–425.

Coombs, R., & Miles, I. (2000). Innovation, measurement and services: The new problematique. In *Innovation Systems in the Service Economy*. Springer, Boston, MA, pp. 85–103.

Davis, J. R., Richard, E. E., & Keeton, K. E. (2015). Open innovation at NASA: A new business model for advancing human health and performance innovations. *Research-Technology Management*, 58(3), 52–58.

Dittrich, K., & Duysters, G. (2007). Networking as a means to strategy change: The case of open innovation in mobile telephony. *Journal of Product Innovation Management*, 24(6), 510–521.

Djellal, F., & Gallouj, F. (2001). Patterns of innovation organisation in service firms: Postal survey results and theoretical models. *Science and Public Policy*, 28(1), 57–67.

Dong, A., & Pourmohamadi, M. (2014). Knowledge matching in the technology outsourcing context of online innovation intermediaries. *Technology Analysis & Strategic Management*, 26(6), 655–668.

Dörner, N., Gassmann, O., & Gebauer, H. (2011). Service innovation: Why is it so difficult to accomplish?. *Journal of Business Strategy*, 32(3), 37–46.

Drejer, I. (2004). Identifying innovation in surveys of services: A Schumpeterian perspective. *Research Policy*, 33(3), 551–562.

Du Chatenier, E., Verstegen, J. A., Biemans, H. J., Mulder, M., & Omta, O. (2009). The challenges of collaborative knowledge creation in open innovation teams. *Human Resource Development Review*, 8(3), 350–381.

Edvardsson, B., & Olsson, J. (1996). Key concepts for new service development. *Service Industries Journal*, 16(2), 140–164.

Enkel, E., Gassmann, O., & Chesbrough, H. (2009). Open R&D and open innovation: Exploring the phenomenon. *R&d Management*, 39(4), 311–316.

Evangelista, R. (2000). Sectoral patterns of technological change in services. *Economics of Innovation and New Technology*, 9(3), 183–222.

Filieri, R. (2013). Consumer co-creation and new product development: A case study in the food industry. *Marketing Intelligence & Planning*, 31(1), 40–53.

Fitzsimmons, J., & Fitzsimmons, M. J. (1999). *New Service Development: Creating Memorable Experiences*. Thousand Oaks, California: Sage Publications.

Foroughi, A., Buang, N. A., Senik, Z. C., Hajmirsadeghi, R. S., & Bagheri, M. M. (2015). The role of open service innovation in enhancing business performance: The moderating effects of competitive intensity. *Current Science*, 109(4), 691–698.

Gadrey, J., Gallouj, F., & Weinstein, O. (1995). New modes of innovation: How services benefit industry. *International Journal of Service Industry Management*, 6(3), 4–16.

Gallouj, F. (2002). *Innovation in the Service Economy: The New Wealth of Nations*. Cheltenham, UK: Edward Elgar Publishing.

Gallouj, F., & Savona, M. (2010). Towards a theory of innovation in services: A state of the art. In: F. Gallouj and F. Djellal (eds.), *The Handbook of Innovation and Services: A Multi- Disciplinary Perspective*, pp. 27–48, Cheltenham, UK.

Gallouj, F., & Savona, M. (2009). Innovation in services: A review of the debate and a research agenda. *Journal of Evolutionary Economics*, 19(2), 149.

Gianiodis, P. T., Ettlie, J. E., & Urbina, J. J. (2014). Open service innovation in the global banking industry: Inside–out versus outside–in strategies. *Academy of Management Perspectives*, 28(1), 76–91.

Grøtnes, E. (2009). Standardization as open innovation: Two cases from the mobile industry. *Information Technology & People*, 22(4), 367–381.

Howells, J. (2006). Intermediation and the role of intermediaries in innovation. *Research Policy*, 35(5), 715–728.

Levine, S. S., & Prietula, M. J. (2013). Open collaboration for innovation: Principles and performance. *Organization Science*, 25(5), 1414–1433.

Loukis, E., Charalabidis, Y., & Androutsopoulou, A. (2017). Promoting open innovation in the public sector through social media monitoring. *Government Information Quarterly*, 34(1), 99–109.

Love, J. H., Roper, S., & Bryson, J. R. (2011). *Openness, knowledge, innovation and growth in UK business services. Research Policy,* 40(10), 1438–1452.

Lusch, R. F., & Vargo, S. L. (2006). Service-dominant logic: Reactions, reflections and refinements. *Marketing Theory*, 6(3), 281–288.

Lusch, R. F., Vargo, S. L., & Wessels, G. (2008). Toward a conceptual foundation for service science: Contributions from service-dominant logic. *IBM Systems Journal*, 47(1), 5–14.

Lyytinen, K., & Rose, G. M. (2003). The disruptive nature of information technology innovations: The case of internet computing in systems development organizations. *MIS Quarterly*, 27(4), 557–596.

Marabelli, M., & Newell, S. (2014). Knowing, power and materiality: A critical review and reconceptualization of absorptive capacity. *International Journal of Management Reviews*, 16(4), 479–499.

Mention, A. L., & Martovoy, A. (2013). *Open And Collaborative Innovation In Banking services: Evidence from Luxembourg*. Public Research Centre Henri Tudor, Luxembourg-Kirchberg.

Martovoy, A., & Mention, A. L. (2016). Patterns of new service development processes in banking. *International Journal of Bank Marketing*, 34(1), 62–77.

Martovoy, A., Mention, A. L., & Torkkeli, M. (2015). Inbound open innovation in financial services. *Journal of Technology Management & Innovation*, 10(1), 117–131.

Mele, C., & Russo-Spena, T. (2017). Innovating as a texture of practices. In *Innovating in Practice*. Springer, Cham., pp. 113–137.

Mergel, I., & Desouza, K. C. (2013). Implementing open innovation in the public sector: The case of Challenge. gov. *Public Administration Review*, 73(6), 882–890.

Metters, R., & Marucheck, A. (2007). Service management — Academic issues and scholarly reflections from operations management researchers. *Decision Sciences*, *38*(2), 195–214.

Miles, I. (2000). Services innovation: Coming of age in the knowledge-based economy. *International Journal of Innovation Management*, 4(04), 371–389.

Mina, A., Bascavusoglu-Moreau, E., & Hughes, A. (2014). Open service innovation and the firm's search for external knowledge. *Research Policy*, 43(5), 853–866.

Morrar, R. (2014). Innovation in services: A literature review. *Technology Innovation Management Review*, 4(4), 6–14.

OECD (2000). Employment in the service economy: A reassessment. *OECD Employment Outlook 2001*, Chapter 3. OECD, Paris, pp. 89–128.

Oke, A. (2007). Innovation types and innovation management practices in service companies. *International Journal of Operations & Production Management*, 27(6), 564–587.

Oliveira, P., & von Hippel, E. (2011). Users as service innovators: The case of banking services. *Research Policy*, 40(6), 806–818.

Pedrosa, A., Välling, M., & Boyd, B. (2013). Knowledge related activities in open innovation: Managers' characteristics and practices. *International Journal of Technology Management* 12, 61(3/4), 254–273.

Piller, F., & Fredberg, T. (2009). *The Paradox of Strong and Weak Ties*. Working Paper RWTH Aachen University and Chalmers University, Aachen and Gothenburg.

Rai, A., & Sambamurthy, V. (2006). Editorial notes — The growth of interest in services management: Opportunities for information systems scholars. *Information Systems Research*, 17(4), 327–331.

Ramaswamy, V., & Ozcan, K. (2018). What is co-creation? An interactional creation framework and its implications for value creation. *Journal of Business Research*, 84, 196–205.

Randhawa, K., Wilden, R., & Hohberger, J. (2016). A bibliometric review of open innovation: Setting a research agenda. *Journal of Product Innovation Management*, 33(6), 750–772.

Salter, A., & Tether, B. S. (2006). Innovation in services. *Through the Looking Glass of Innovation Studies*. Tanaka Business School, Imperial College, London.

Savino, T., Messeni Petruzzelli, A., & Albino, V. (2017). Search and recombination process to innovate: A review of the empirical evidence and a research agenda. *International Journal of Management Reviews*, 19(1), 54–75.

Shamseer, L., Moher, D., Clarke, M., Ghersi, D., Liberati, A., Petticrew, M., Shekelle, P., & Stewart, L. A. (2015). Preferred reporting items for systematic review and meta-analysis protocols (PRISMA-P) 2015: Elaboration and explanation. *BMJ*, 349, 7647.

Sundbo, J. (1997). Management of innovation in services. *Service Industries Journal*, 17(3), 432–455.

Temel, S., Mention, A. L., & Torkkeli, M. (2013). The impact of cooperation on firms' innovation propensity in emerging economies. *Journal of Technology Management & Innovation*, 8(1), 54–64.

Tether, B. S. (2005). Do services innovate (differently)? Insights from the European innobarometer survey. *Industry & Innovation*, 12(2), 153–184.

Van de Vrande, V., De Jong, J. P., Vanhaverbeke, W., & De Rochemont, M. (2009). Open innovation in SMEs: Trends, motives and management challenges. *Technovation*, 29(6–7), 423–437.

Vargo, S. L., & Lusch, R. F. (2017). Service-dominant logic 2025. *International Journal of Research in Marketing*, 34(1), 46–67.

Vargo, S. L., Maglio, P. P., & Akaka, M. A. (2008). On value and value co-creation: A service systems and service logic perspective. *European Management Journal*, 26(3), 145–152.

Witell, L., Snyder, H., Gustafsson, A., Fombelle, P., & Kristensson, P. (2016). Defining service innovation: A review and synthesis. *Journal of Business Research*, 69(8), 2863–2872.

https://doi.org/10.1142/9789811234491_0003

Chapter 3

Is there a Need for "Open Service Innovation" Term: A Scientometrics Analysis of Open Service Innovation Research Domain

Teemu Santonen

Laurea University of Applied Sciences Vanha maantie 9, 02650 Espoo, Finland

teemu.santonen@laurea.fi

Abstract. Science by definition builds on previous knowledge and therefore the main goal of this study is to systematically evaluate the history and current status of the open service innovation (OSI) research by utilizing popularity-based and impact based scientometrics methods. The results grounded on ISI Web of Science (WoS), Scopus and Google Scholar data sources reveals that OSI research domain is still infancy and so far there are only handful authors and publications which have embraced the OSI term. However, the number of publications and authors focusing on open innovation (OI) in conjointly with services or service innovation (SI) is much greater. Albeit, there is a latent potential for OSI studies, but a need for a new OSI term remains questionable especially when interest towards OI and SI have recently leveled. A clear differences between WoS, Scopus and Google Scholar as a data source was identified. Therefore, the further studies focusing on innovation management execute triangulation in data source selection.

Keywords. Open service innovation; open innovation; service innovation; service; scientometrics; bibliometrics; data triangulation; systematic literature review; Google Scholar; Web of Science; WoS; Scopus; popularity based analysis; actor analysis; impact analysis.

1. Introduction

Science by definition builds on previous knowledge, which evolves over time, refines and develops knowledge and serves as a foundation for further research. Thus, in-depth understanding of scientific knowledge and it's evolution in specific research themes such open service innovation (OSI) is vital. The concept of research paradigm and paradigm shift was popularized by Kuhn (1962). Later Dosi (1982) proposed that continuous innovation can be regarded as an event within a paradigm, whereas discontinuous innovation could be the starting point for a new paradigm. This fundamental idea has then been applied in various academic fields in order to understand the history, current state and future of the given paradigm (Gupta and Bhattacharya, 2004).

In innovation management research, many literature reviews have been based on narratives (McLean, 2005), and to a lesser extent do we see more rigorous research methods such as systematic literature reviews (Becheikh *et al.*, 2006), scientometrics (Larivière *et al.*, 2012), bibliometric analysis (Pritchard, 1969) or informetrics (Nacke, 1979). The terms bibliometrics, scientometrics, and informetrics are used to describe similar and overlapping methodologies which are quantitatively evaluating changes in the output of a scholarly field through time (Hood and Wilson, 2001). For simplicity in this study we are using scientometrics term, which can be defined as the quantitative study of science and technology (Van Raan, 1998). The main goal of this study is to systematically evaluate the history and current status of the OSI research by utilizing scientometrics methods.

2. Prior Scientometric Studies on Open Innovation

In prior literature, there are only few scientometrics studies which have focused on open innovation (OI) research theme (Chesbrough, 2006). According to citations, the most impressive study appears to be "How Open is Innovation?" by Dahlander and Gann (2010) which by combing bibliographic analysis were clarifying the definition of "openness". The main outcome of the study was identification and comparison of four different types of openness. The data source in this study was Thomson's Web of Science (WoS) which at the time of study was known as Web of Knowledge. The data collection was conducted in August 2009, which at the time resulted 701 papers. Due the intense evolution and expansion of OI research, the year 2009 data in today's criteria is limited.

A more recent study by Randhawa, *et al.* (2016) evaluated existing gaps in OI research, and suggested new avenues for future OI research. Instead of WoS, these authors were using Scopus database which has more extensive coverage than WoS (Falagas *et al.*, 2008). Besides relying only on OI search term authors were using also openness, co-creation and crowdsourcing as search terms since they were considered to be closely related to OI concept. After excluding journals without impact factor, filtering duplicates and non-relevant papers, the total publication count was 321 papers. The following three distinct OI research areas were identified: (1) firm-centric aspects of OI, (2) management of OI networks, and (3) role of users and communities in OI. Importantly authors also identified that so far OI in services has not been a particularly popular topic (Chesbrough, 2011) and there is a need to enhance focus on customer co-creation and conceptualize "OSI".

A study by Kovacs *et al.* (2015) combined bibliographic coupling and co-citation analysis and evaluated the network and clusters of OI publications. The data for this study was based on WoS dataset (from 2003 to 2013) including 358 publications, which referred to 11.873 publications and was referred by 2.372 other publication by themselves. The results suggest that OI research is mainly grounded on following four related streams of prior research: (1) Strategic partnering and external sourcing, (2) User-centric innovation, (3) Technology and innovation management, (4) Resource- and knowledge-based view of the firm. The OI research itself was divided into following seven thematic clusters: (1) The Core of OI, (2) User-Centric Innovation, (3) External Knowledge Sourcing, (4) External Technology Commercialization, (5) Implementation Mechanisms and Tools, (6) OI in Specific Industries and (7) Idea Generation and Idea Competitions.

There are also few other studies which have presented frameworks for OI studies and suggested avenues for future research including Hossain *et al.* (2016) study which was mainly grounded on WoS dataset including 293 papers, but conducted content analysis only on most important articles. A simple descriptive quantitative analysis relating frequencies of disciplines, location of authors (country and region), journals and keywords was also conducted. As a result authors suggested a framework for future research, but comparing to Randhawa *et al.* (2016), the analysis is clearly more superficial and mainly based on narrative approach instead of rigorous empirical analysis.

Partially by the same authors as above, Hossain and Anees-ur-Rehman (2016) conducted another study based on WoS database including 411 OI publications. The approach appeared to somewhat similar as in Hossain *et al.* (2016) but the analysis was conducted a bit more detailed level including distribution number of publications, articles by journal, authorship by country, number of authors, authorships in terms of (a) single-country vs. multiple-country, (b) inter- vs. intra-regional, (c) developed vs. developing countries, (d) data source region, (e) unit of analysis, (f) research design, (g) analytic techniques, (h) industry, and (i) research streams based on keywords. Respectively to Hossain *et al.* (2015) and on the contrary to Randhawa *et al.* (2016), also this study is more or less grounded on descriptive approach instead of more rigorous empirical analysis.

Finally there are also two other bibliographical studies which have evaluated OI literature but in combination with an additional specific theme. Medeiros *et al.* (2016) evaluated OI in context of agrifood chain by using WoS, Scopus and Science Direct databases, but ended up founding only 37 publications. Hossain (2015) studied OI literature in context of small and medium-sized enterprises (SMEs) and grounded data collection on WoS, Scopus and Google Scholar. Also his data collection resulted very limited number of publications (61 relevant publications). As indicated already above, Randhawa *et al.* (2016) study also suggested that services have not been a popular topic among OI scholars. Therefore it is argued that when OI is linked to more specific research theme such service innovation (SI), the number of available publication will significantly diminish comparing to total OI body of knowledge.

3. Research Methodology

3.1. *Definition of overall research framework and objectives of this study*

Recently, Santonen and Conn (2015a) illustrated a comprehensive framework for classifying various types and combinations of scientometric studies when studying actors and/or contents within a particular research theme (see Figure 1).

The framework includes the following three main research approaches: (1) "popularity-based" studies (Choi *et al.*, 2011), (2) impact based studies grounded on citation/co-citation studies (Pilkington and Meredith, 2009), and (3) social network analysis (SNA) studies (Wasserman and Faust, 1994). "Popularity-based studies" are typically analyzing frequency of people, keywords or other related meta-terms, which have been derived from the context of the research publication. "Citation and co-citation studies" focuses on the importance and impact of the people or topics by evaluating how much (popularity approach) or by whom (network approach) the particular study is cited (Pilkington and Meredith, 2009). "SNA-based studies" instead focus on the relationships via publications, which most typically are based on co-authorship (e.g., Su and Lee, 2012) or keywords (Yi and Choi, 2011).

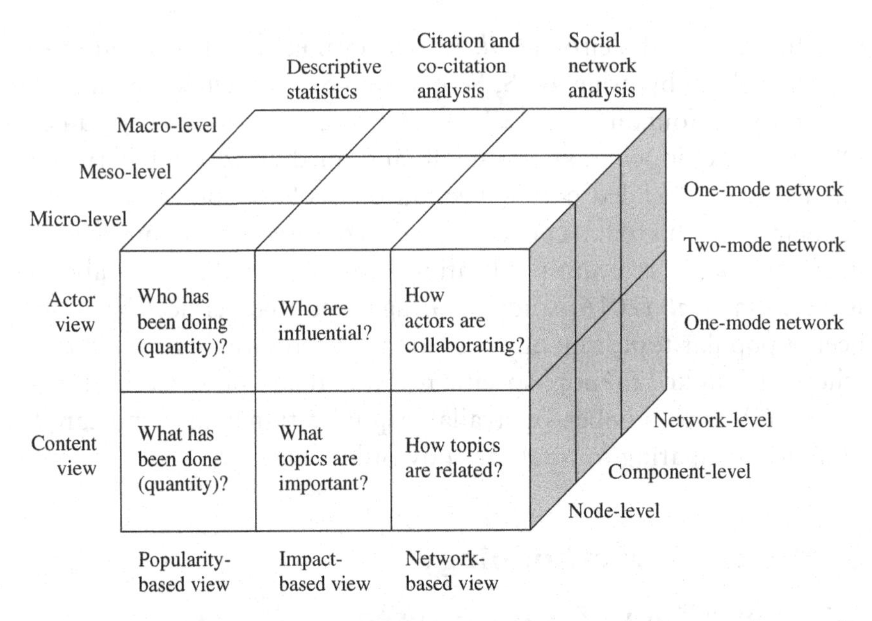

Figure 1. Comprehensive framework for classifying Scientometric studies, modified by author from Santonen and Conn (2015a).

All the above research viewpoints can be applied at the micro-level (e.g., individual authors or keywords) or the meso-level (e.g., universities that the authors represent or combined research theme which includes multiple subtopics) or at the macro-level (e.g., countries or research paradigm) (Gupta and Bhattacharya, 2004). Respectively, SNA studies can focus on node-, component or network level analysis. In SNA studies a node can refer to any kind of actor within a network (e.g., an author, country, keyword) whereas a component is a combination of nodes which are directly or indirectly connected by at least one connection. In an extreme case a network can consist only one component and then this so called main component (i.e., the largest component) is the same as the whole network. In a network level analysis, the unit of analysis is the whole network. There are two types of networks: one- and two-mode networks. One-mode network refers to a distinct set of entities (e.g., authors or keyword), whereas two-mode network as the name indicates includes two (or more) set of entities (e.g., author and keyword). One-mode studies can for example evaluate how authors are collaborating, while two-mode studies can reveal which authors have focused on certain research topics. Finally, scientometrics studies can either focus on a snapshot of given time or include temporal data which can better reveal the evolution of the research domain.

The goal of this study is to analyze the current status and the historical evolution of "OSI" research theme by applying (1) popularity-based bibliometric analysis (Choi *et al.*, 2011), and (2) citation analysis as suggested by Santonen's and Conn's (2015a) comprehensive scientometrics framework. The methodological framework, data source selection and data collection and analysis strategies are based on the authors prior publications (Santonen, 2018; Santonen *et al.*, 2018; Vilko & Santonen, 2017; Santonen & Conn, 2016).

3.2. *Selecting the data sources*

Triangulation is derived from navigation and military strategies (Smith, 1975, p. 273) and in short can be defined as (Denzin, 1978, p. 291) "the combination of methodologies in the study of the same phenomenon". There are many possible approaches for triangulation including the data triangulation — gathering data at different data sources.

Some studies have suggested that Scopus has more extensive coverage than ISI WoS (Falagas *et al.*, 2008). Therefore, the data for this study was collected from both of these databases in order to increase the robustness of our data analysis and reveal possible differences in OSI research domain.

Importantly, in some studies Google Scholar has been suggested as an alternative or complementary resource to the Scopus and WoS since it includes also the local contents, papers in low impact journals, popular scientific literature, and unpublished reports and teaching supporting materials (Aguillo, 2011). Google Scholar has also suggested to have a better coverage of conference proceedings (Meho and Yang, 2007) which have been recognized as an alternative but relevant and important knowledge source although also criticized by many authors (e.g., Lisée *et al.*, 2008). In addition Scopus and WoS have more limited coverage in the management studies than Google Scholar (Harzing and Van Der Wal, 2009; Mingers and Lipitakis, 2010). Since "OI" and "SI" studies which are the underlying foundation for "OSI" research are typically conducted by management scholars, it is possible that they are not fully able to detect the research trends as good as more extensive Google Scholar.

However, albeit having more extensive coverage, Google Scholar has also some weaknesses. Prior studies have argued that Google Scholar is more inadequate than commercial scientific databases (Falagas *et al.*, 2008) and several problems and shortcomings have been detected (Aguillo, 2011). Even if Google Scholar has shown substantial expansion over the years (De Winter *et al.*, 2014), the quality control in Google Scholar is not as good as in commercial scientific databases.

In spite of the above quality differences, according to data triangulation principles it is important to execute at least some of the analysis in all three databases but taking into account the possible error sources such as possibility to target the search queries.

Both Scopus and WoS provides various options to target search in a particular database fields. In order to make the results as comparable as possible, in the case of WoS topic search (search from title, abstract, author keywords and keywords plus which consist of words and phrases harvested from the titles of the cited articles) and in the case of Scopus (title, keywords, or abstract) was applied.

Targeting searches in Google Scholar is more problematic than in Scopus and WoS. The only options targeting searches at field level are "only in title" or "anywhere in article". The both these search options are not comparable with Scopus or WoS. Targeting search to "title" is extremely limited and excludes abstract and keywords which are available in Scopus and WoS. "Anywhere in article" is too extensive, since it includes also additional fields beyond WoS and Scopus. Therefore, the absolute number of publications are not directly comparable with Scopus and WoS, but the relative frequency comparison within Google Scholar results could be compared to relative Scopus and WoS findings.

3.3. *Data collection process*

The unit of analysis in this study is a scientific publication which topically focuses on "OSI". Huizingh (2011) argued that the use of outside knowledge as a driver for innovation, did not originate from the OI paradigm but from a rich stream of various research paradigms. He also predicted that *open innovation is on its way to become innovation since it will become fully integrated in innovation management practices.*

Recently, by using innovation management conference publications as a data source, Santonen and Conn (2016) provided an empirical support for Huizingh's predictions since they identified that innovation conference publications TOP25 keyword list shared 40% of common keywords between "OI" and "innovation" publications. Furthermore, another study by Santonen and Conn (2015b) suggested that innovation management conference publication research topic frequency distribution is somewhat similar in different years although subtle incremental change over the years can be detected. As a result is it assumed that also OSI research will be partially linked to "SI" and "OI" research streams and the changes will occur incrementally. Therefore in order to predict where "OSI" research stream would or could evolve over time, it is important to understand what has happened in "SI" and "OI" research streams, since over time we expect them to emerge.

As a result in order to reveal the relevant "OSI" contributions, the search criteria must be based on all "OI studies which are focusing on services and SI" instead of just relying on the combined "OSI" term. The independent "OI" search term will reveal all the OI studies whereas independent "SI" will reveal all the SI studies. Furthermore, "OI" search term can be combined separately with "SI" (OI+SI) and "service" (OI+S) to reveal all the possible OI studies which have shown interest towards services. The separation between "SI" and "service" was considered important since "service" term is less bounding and could therefore reveal additional insight. Finally, the "OSI" term will be used as indicator of this research domain core. In Table 1, the search term queries and their abbreviation are presented.

Table 1. Definition and abbreviations of search term alternatives.

Search term(s)	Abbreviation
Open innovation	OI
Service innovation	SI
Open innovation AND service	OI + S
Open innovation AND service Innovation	OI + SI
Open service innovation	OSI

Finally, besides innovation management scholars, OSI might have gained an interest among multiple other disciplines, especially in the other fields of management such as marketing, strategy, economics, but also in decision sciences, social sciences, engineering and computer science. Therefore in data filtering, we followed similar strategy as Zhao and Zhu (2012) and did not limited the searchers by the academic disciplines. This approach has also been common in many other scientometric studies. Timewise, the searches are limited to years 2003 to 2016, since the Chesbrough's original book was published in 2003 and 2016 is the last full year available at the time of the data collection.

4. Results — Popularity-based Analysis at Content Level

4.1. *Popularity based view at cumulative content level*

In Table 2, comparison between cumulative number of publications in year 2016 in Scopus, WoS and Google Scholar is presented.

As suggested by prior studies (Falagas *et al.*, 2008), also in this comparison reveals that Scopus has more extensive coverage than WoS. During the 2003–2016 WoS search resulted 2.233 publications for OI, 1.486 for SI and 378 for OI+S whereas Scopus resulted 3.272 publications for OI (46.5% more than WoS), 2.496 for SI (68.0% more than WoS) and 631 for OI+S (66.9% more than

Table 2. OI, SI, OI+SI and OSI publications cumulative popularity comparison between Scopus, WoS and Google Scholar in year 2016.

	WoS	Scopus	Google Scholar
OI	2,233	3,272	64,895
SI	1,486	2,496	42,325
OI+S	378	631	39,797
OI+SI	41	76	5,409
OSI	15	26	468

WoS) publications. So far the OSI term has gained only modest popularity among scholars. In Scopus there are only 26 publications and in WoS only 15 publications. OI+SI search resulted only slightly better result (76 publications from Scopus and 41 publications from WoS). Also this outcome can be considered modest. As assumed the number of publications for each search query (OI, SI, OI+SI, OSI) is significantly more substantial in the case of Google Scholar. By the year 2016 there was 64.895 OI, 42.325 SI and 39.797 OI+S, 5.409 OI+SI and 468 OSI publications.

In Table 3, variables relative share to OSI, OI and SI is presented to compare the search result balance between WoS, Scopus and Google Scholar.

As indicated in Table 3 most of the search results are relatively in the same magnitude level between WoS, Scopus and Google Scholar excluding the number of OI+SI and OI+S publications in Google Scholar. The relative share of OI+S/OSI is about 25 in WoS and Scopus but for Google Scholar as high as 85. In the case of OI+SI/OSI, WoS and Scopus relative share is about 3 and in the case of Google Scholar almost 12. Also in the case of OI studies, results between WoS and Scopus are also differing a bit more. In WoS OI studies are about 150 times more popular and in Scopus about 125 times more popular. In this case Google Scholar is in middle ground (140 times more popular).

Importantly, in all three databases relatively OSI publications are covering less than 1% from OI studies and about 1% from SI

Table 3. Relative comparison of OI, SI, OI+SI and OSI between WoS, Scopus and Google Scholar in year 2016.

Relative to	OI	SI	OI+S	OI+SI	OSI
OSI — WoS	148.87	99.07	25.20	2.73	1
OSI — Scopus	125.85	96.00	24.27	2.92	1
OSI — GS	138.66	90.44	85.04	11.56	1
Mean (all)	137.79	95.17	44.84	5.74	1

studies. Admittedly this verifies that OSI as a term is by any means infancy, but also that SI as a term is not extensively incorporated with OI studies especially in WoS and Scopus. There are only handful of OSI studies and also only few more OI studies which define SI as a term, but clearly more OI studies focusing on service(s).

This finding is important and can indicate that OI and SI research domains are following different traditions and definitions. As suggested by Santonen *et al.* (2016) previous research has emphasized the need for a more parsimonious understanding of innovation typology, since otherwise it will lead to competition for rivaling types and lack of research efficiency. As a result it is argued that there is a latent potential for OSI studies, but in order to redeem that, the terminology and definitions for OSI needs to be clarified.

4.2. *Evolution of cumulative popularity at content level*

In Figure 2, the evolution of cumulative number of OI, SI and OI+S publications from year 2003 to year 2016 is compared between Scopus and WoS and then Google Scholar in Figure 3.

As a result it appears that SI was actually more popular research topic than OI until year 2010 in WoS and until year 2012 in Scopus. In Google Scholar, the popularity transition in favor for OI took place earliest, in year 2008. In this case the assumption that Google Scholar could reveal the research trends faster than WoS or Scopus was verified. After the transition, OI has been able to growth

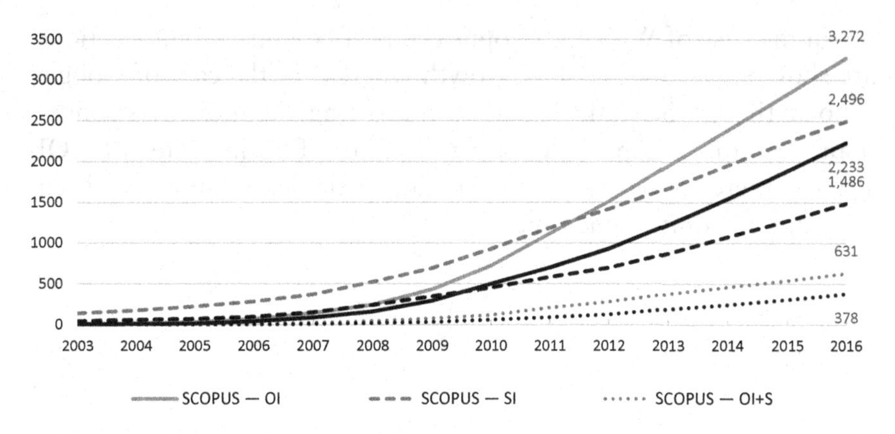

Figure 2. Cumulative popularity — "OI" vs. "SI" vs. "OI + service" — WoS and scopus.

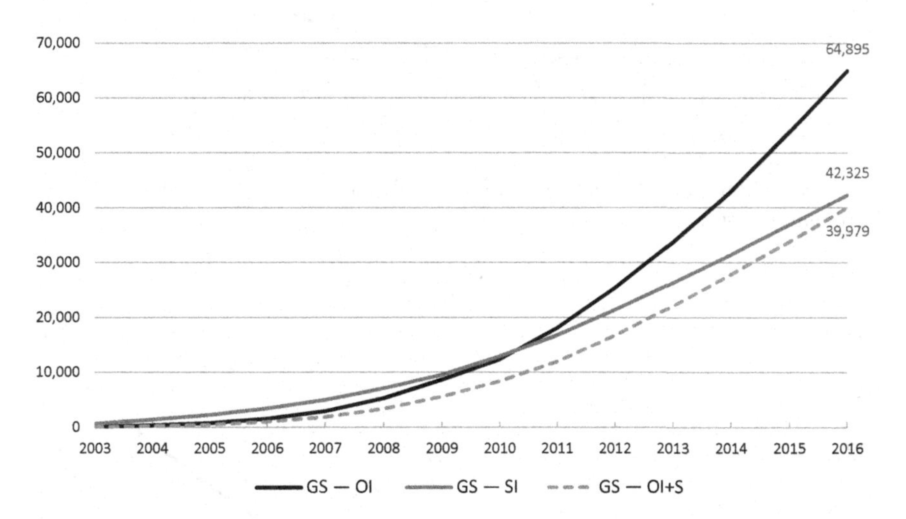

Figure 3. Cumulative popularity — "OI" vs. "SI" vs. "OI + service" — Google Scholar.

stronger especially in Google Scholar and has been able to increase the popularity gap between OI and SI studies.

The evolution of cumulative number of OI+SI and OSI from year 2003 to year 2016 is compared between WoS and Scopus in Figure 4 and in Figure 5 in Google Scholar.

In the case of WoS and Scopus both OI+SI and OSI publications are showing steady but slow growth whereas in the case of Google Scholar the OI+SI studies have been growing significantly stronger than OSI studies especially after year 2010. This indicates that OI scholar interest towards SI has not happened over night but has been emerging for quite some time.

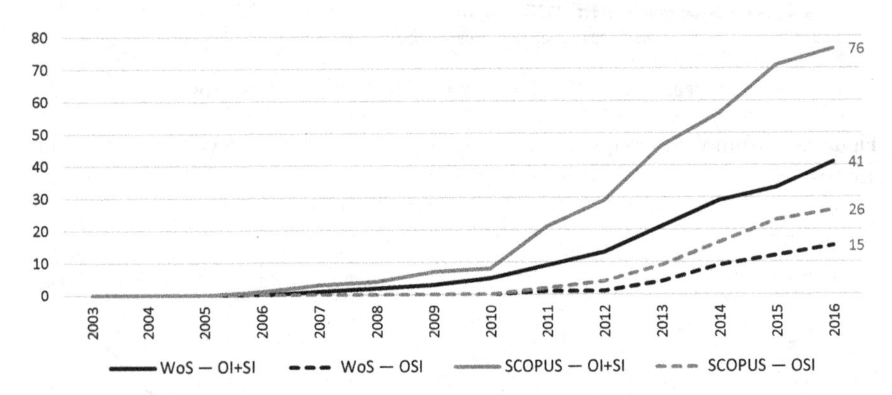

Figure 4. Cumulative popularity — "OI and SI" vs. "open service innovation" — WOS and Scopus.

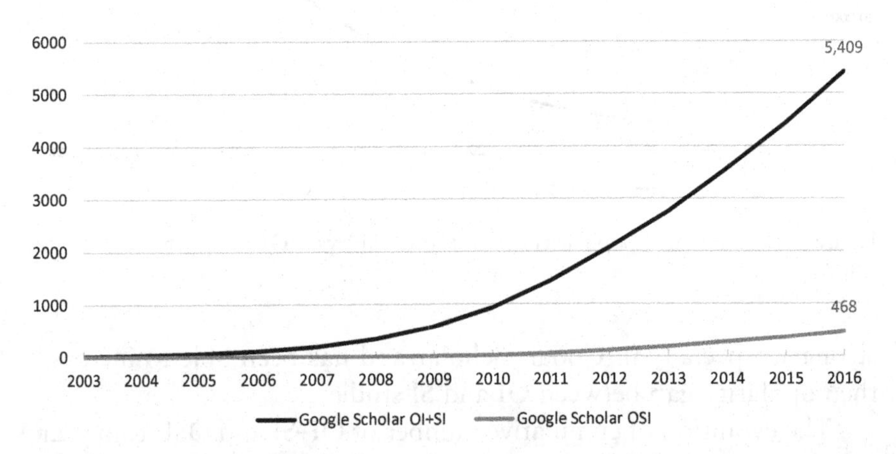

Figure 5. Cumulative popularity — "OI" vs. "SI" vs. "OI + service" — Google Scholar.

4.3. *Evolution of popularity annually at content level*

Comparing to cumulative popularity, the annual amount of publications can reveal faster the possible popularity changes and trends of the given research domain. Therefore, in Figure 6 evolution of annual popularity in WoS and Scopus is compared between OI, SI and OI+S.

According to Scopus search results, a clear popularity boost in OI research was occurring from 2008 to 2011 but after that the growth has been clearly leveling and currently in year 445 OI studies are published. In the case of WoS, the popularity boost is more gentle and currently about one hundred studies are published less. In all from year 2015 to year 2016 interest towards OI has been leveling. Until year 2009 the popularity of SI studies were in head-to-head competition with OI studies, but after that SI studies showed only minor growth especially in the case of WoS. From year 2015 to year 2016 Scopus is indicating declining while WoS is still showing minor growth and therefore the gap between WoS and Scopus seems

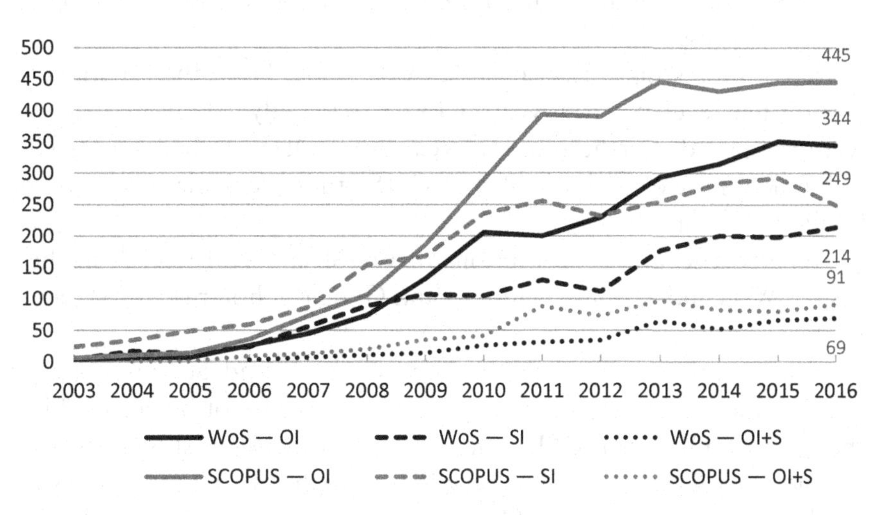

Figure 6. Annual popularity — "OI" vs. "SI" vs. "OI + service" — WOS and Scopus.

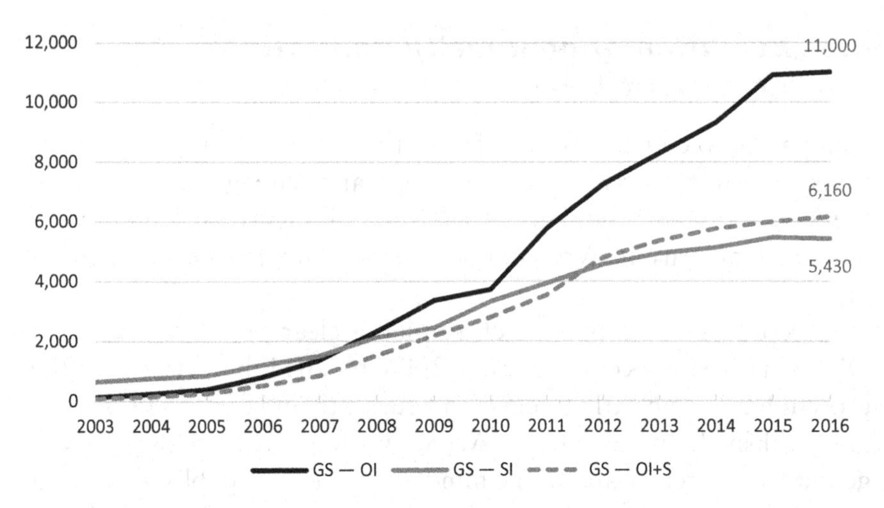

Figure 7. Annual popularity — "OI" vs. "SI" vs. "OI + service" — Google Scholar.

to be closing. In year 2016, there are 249 SI studies in Scopus and 214 studies in WoS. Studies focusing on OI+S indicate more modest growth are not reaching hundred studies in a year.

In Figure 7 evolution of OI, SI and OI+S publications in Google Scholar is presented.

Two interesting observations can be made. First, the OI studies growth has leveled. From year 2015 to 2016 only 100 more studies were published. Furthermore, in year 2012 OI+S studied surpassed SI studies. However, both SI and OI+S studies are also indicating leveling interest.

Finally also annual evolution for OI+SI and OSI studies in the case of WoS and Scopus (Figure 8) and Google Scholar (Figure 9) are analyzed.

As a result the comparison between WoS and Scopus reveals more random interest and clear growth trends cannot be detected in the case of OI+SI or OSI, whereas Google Scholar shows steady growth for OI+SI studies but very low growth for OSI studies. As a result the gap between OI+SI and OSI studies in Google Scholar is expected to expand even more in the future.

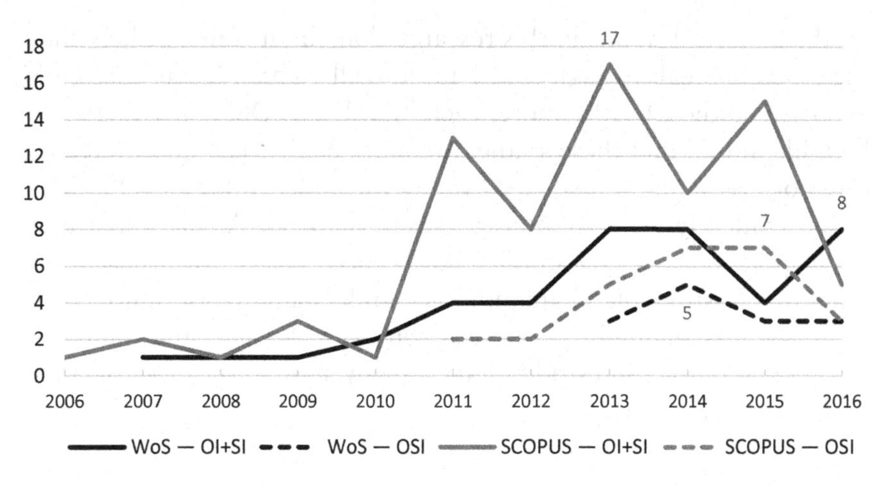

Figure 8. Annual popularity — "OI AND SI" vs. "OSI"– WOS and Scopus.

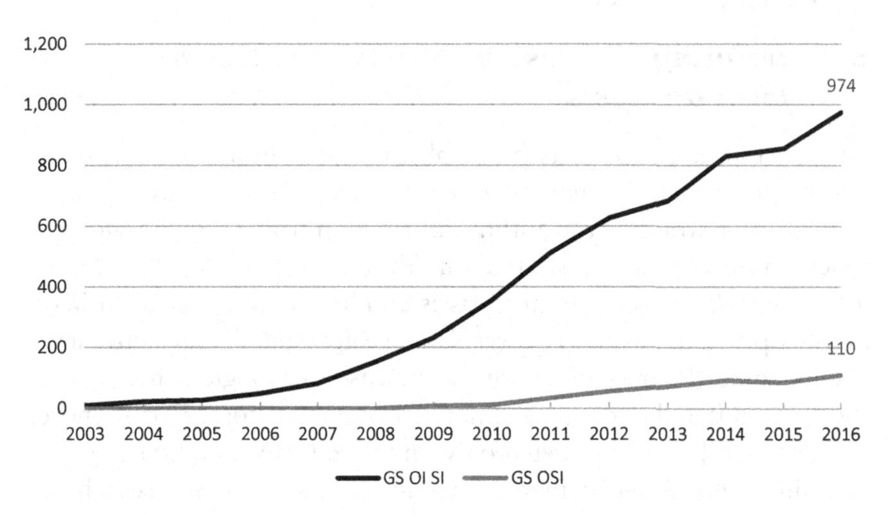

Figure 9. Annual popularity — "OI AND SI" vs. "OSI" — Google Scholar.

4.4. *Discussion*

In summary, when evaluating publications at annual level, it is concluded that the popularity of OSI as a term is not expected to increase substantially in coming years unless breakthrough articles are published, which will bring OSI in the spotlight. However, based on the historical trends in this research domain, it is more likely that the possible breakthrough study/studies will be branded as an OI + S or OI+SI studies. Furthermore, even branded as OSI study there is a high likelihood that these studies are more likely to be interpreted as OI+S or OI+SI studies, since currently the added value for OSI definition evidently remains blurry among scholars. Albeit, there is a latent potential for OSI studies, but is there a need for a new term remains questionable. Parsimonious understanding of innovation typology was highlighted in Santonen *et al.* (2016) study and OI studies and innovation studies have already started to merge together (Santonen and Conn, 2016).

5. Results — Popularity-based Analysis at Actor Level

5.1. *Defining the research focus for actor level analysis*

Besides counting popularity by number of publications, also the number of authors can be regarded as a good indicator of research domain popularity. Extracting the author information from Google Scholar is much more complex task than in the case of WoS and Scopus. Therefore, the actor level analysis is conducted only based on WoS and Scopus databases. As a result it is highlighted that actor level results could differ significantly in the case of Google Scholar since they have differed in the case of number of publications. Furthermore, the actor level will be targeted only on OI+S, OI+SI and OSI research domains, which are considered to be in the heart of OSI research.

5.2. *Evolution of OI+S cumulative and annual popularity at actor level*

First, both the cumulative amount of different authors and the amount new authors joining annually relating OI+S search results is compared between Scopus and WoS in Figure 10. Due the large size

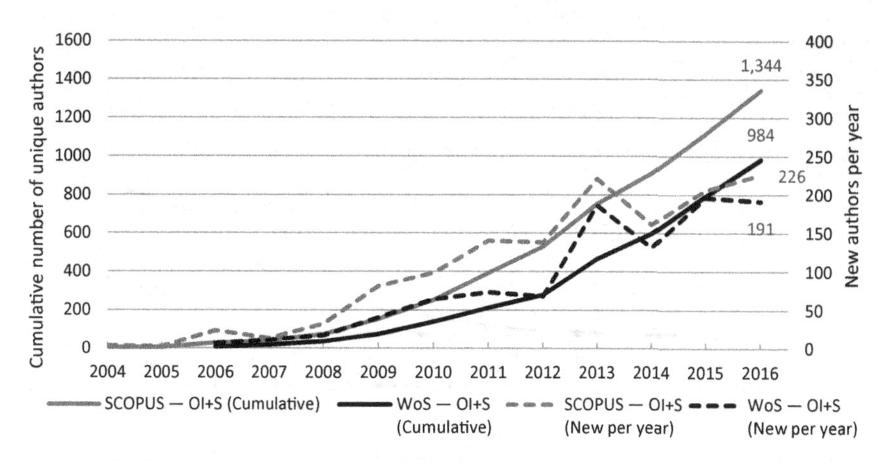

Figure 10. Cumulative number of different authors vs. new authors per year — "OI AND service" — WOS vs. Scopus.

difference, the left hand side vertical axis is measuring the evolution of cumulative number of different authors, whereas right hand side vertical axis is measuring the number of new authors in each year. By the year 2016, the number of different authors writing about OI+S is 1.344 in the Scopus and 984 authors in the WoS. Respectively to prior content based view results, the annually new authors measure is indicating somewhat leveling interest towards OI+S topic. Currently about 200 new authors become a part of OI+S research domain.

5.3. *Evolution of OI+SI and OSI cumulative and annual popularity at actor level*

Second, in Figure 11 the comparison between cumulative amount of different authors relating OI+SI and OSI search results are presented. Especially the OSI community of authors is very limited. Scopus database identifies 61 authors whereas WoS only 38. The number of authors focusing on OI+SI is also modest (Scopus 146 and WoS 105).

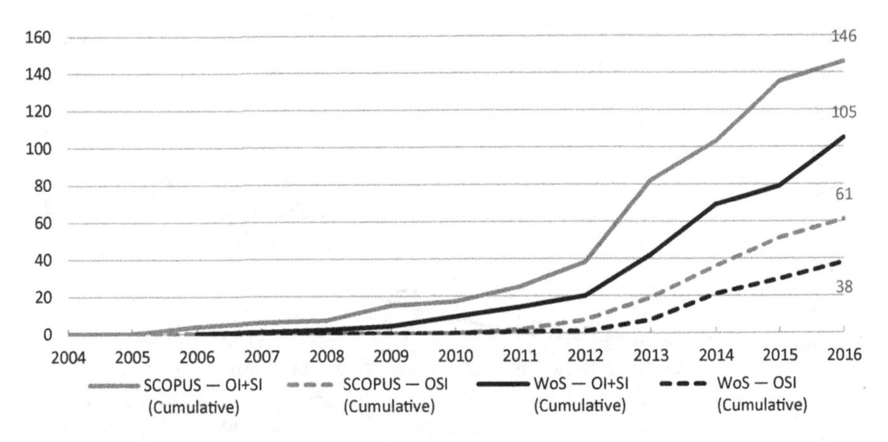

Figure 11. Cumulative number of different authors — "OI and SI" vs. "OSI" — WoS vs. Scopus.

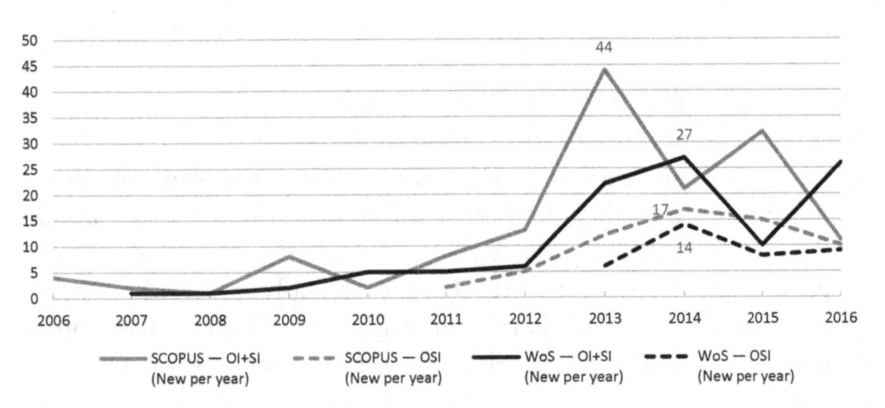

Figure 12. New authors per year — "OI AND SI" vs. "OSI" — WoS vs. Scopus.

Furthermore, in Figure 12 the number of new authors annually relating OI+SI and OSI are also compared between Scopus and WoS. In the case of Scopus a clear decreasing leveling is observed for OI+SI since so far the highest peak has been in years 2013 which included 44 new authors while in year 2016 there was only 11 authors. The highest peak in WoS was in year 2014 (27 new authors) but in year 2015 only 10 authors. On the contrary to Scopus, WoS recovered in year 2016 to 26 new authors. After the peaking year 2014, OSI results are indicating decreasing growth pattern for both databases.

5.4. *Discussion*

The popularity based result from actor point of view verifies the prior content level observations. However, the problem appears to be bigger in the case of tempting new authors, since the decaling (OI+SI and OSI) and leveling (OI+S) trends were detected. In his book, Moore (1996) defined four distinct stages for the lifecycle of an ecosystem: pioneering, expansion, authority, and renewal or death. Evidently, if OSI research domain is considered as an ecosystem, it is clearly still in the pioneering stage, while so far OI has so far experienced expansion stage and achieved the authority stage by providing a compelling vision that encourages other actors to improve the ecosystem. However, there are now first indications that OI should enter to renewal stage and start bringing new separating ideas or there is a threat it will merge and become as innovation research. This might be problematic also for OSI domain which is still struggling to establish itself as own sub domain in innovation management research.

6. Results — Impact-based Analysis

6.1. *Impact based view — Content level*

In Figure 13, cumulative citations of OI+S are compared between Scopus and WoS databases.

The first citations on OI+S publications are starting from year 2005 in Scopus and from year 2008 in WoS. Until year 2016 OI+S publications in Scopus have generated 3.496 citations and in WoS 2.306 citations. After the first few years, the balance favor relating annual citations for Scopus is reducing. By the year 2016 publications in Scopus indicated 1.52 more citations than WoS.

In Figure 14, annual citations of OI+S are compared between Scopus and WoS. This comparison result is indicating steady citation growth in both databases. It appears that OI+S publications in WoS are focusing on higher quality than in Scopus since in the case of

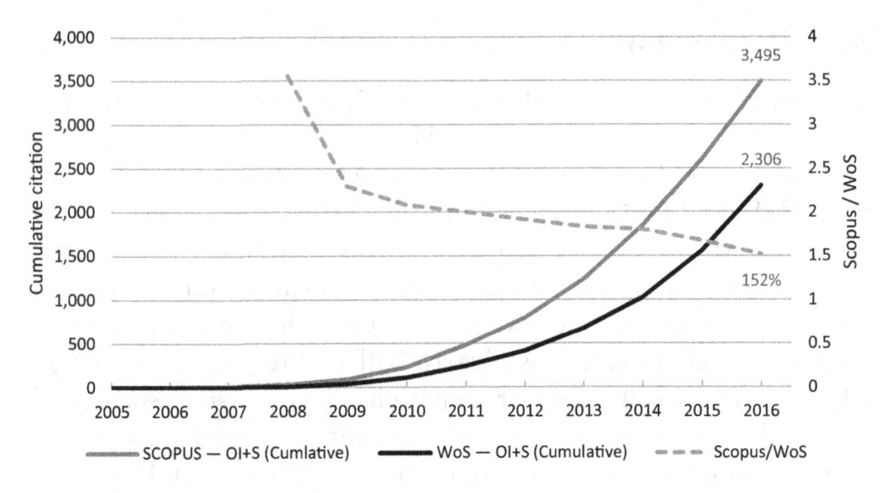

Figure 13. Cumulative citations — "OI + Service" — Scopus vs. WOS.

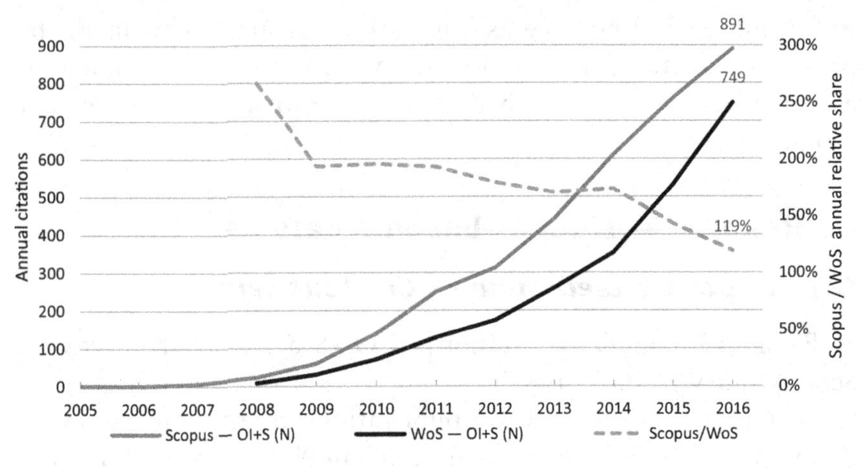

Figure 14. Annual citations — "OI + service" — Scopus vs. WOS.

annual citations comparison Scopus publications has only 1.19 more citations than WoS.

Respectively to prior popularity based analysis, in Figure 15, cumulative citations between OI+SI and OSI are also compared between Scopus and WoS.

In the case of OI+SI, the first citations appears in Scopus in year 2008 and three years later in WoS (year 2011). In the case of OSI the

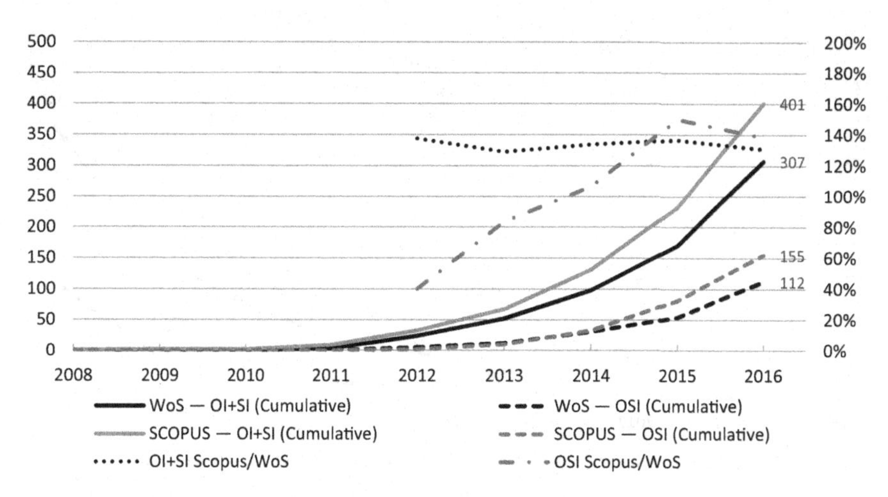

Figure 15. Cumulative citations — "OI + SI" and "OSI" — Scopus vs. WoS.

first citations appears in 2012 both in Scopus and WoS. Cumulatively in Scopus there are 401 OI+SI and 155 OSI citations, whereas in WoS there are 307 OI+SI and 112 OSI citations. Between 2012 and 2013, OSI publications has more citations in WoS than in the Scopus. However, the total amount of citation is very low in both cases (WoS 2012 = 5, 2013 = 12; Scopus 2012 = 2, 2013 = 10). By the 2016 Scopus OI+SI publications has 1.31 more citations than WoS and 1.38 more OSI publications.

In Figure 16, annual citations between OI+SI and OSI are also compared.

These results are showing steady growth. In year 2016, OI+SI studies are generating 169 citation and OSI studies 74 citations in Scopus. In WoS, there are 137 OI+SI citations and 58 OSI citations.

As presented in Figure 17, the annual citation balance between Scopus and WoS relating OI+SI and OSI is also mixed. Nevertheless the prior findings that Scopus has more citations is also validated in this case.

In Table 4, a comparison between cumulative number of publications and citations in year 2016 are compared in order to reveal how many citations one publications is generating. On the average one OI

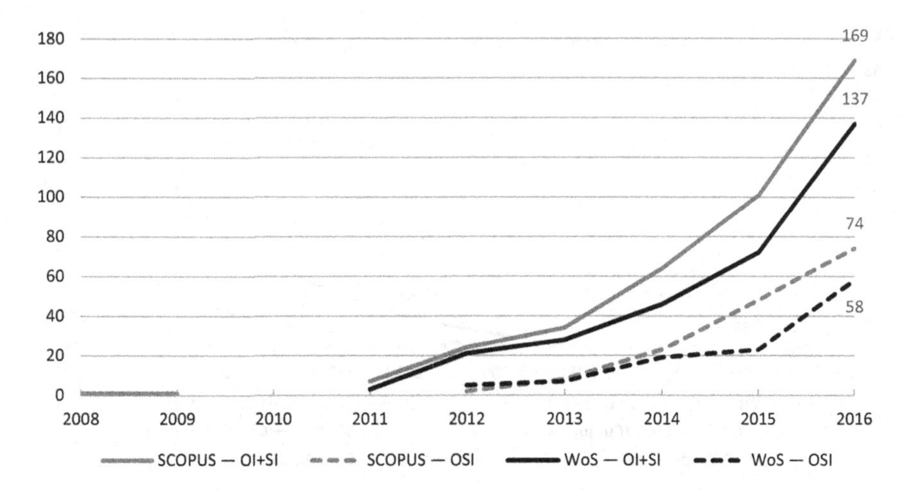

Figure 16. Annual citations — "OI + SI" vs. "OSI" — Scopus vs. WOS.

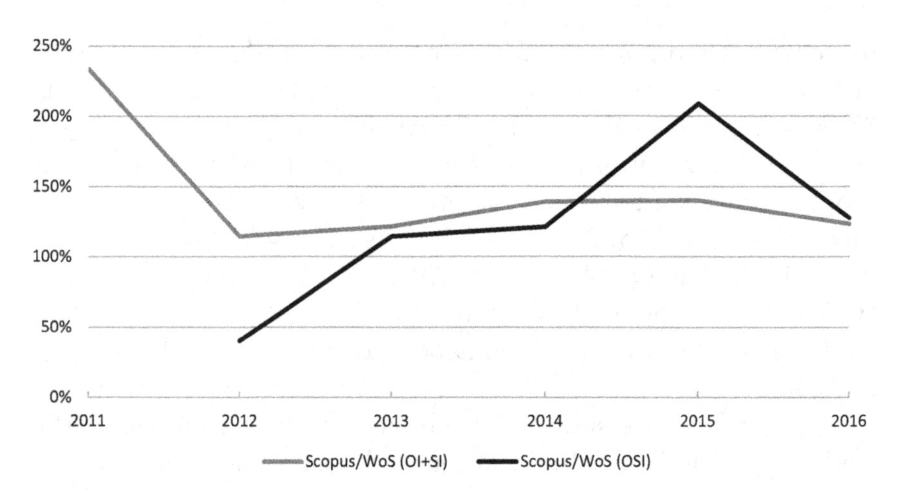

Figure 17. Annual citations balance between "OI + SI" vs. "OSI" in the case of Scopus vs. WoS.

Table 4. Publications/citations ratio.

	WoS OI+S	WoS OI+SI	WoS OSI	SCOP. OI+S	SCOP. OI+SI	SCOP. OSI
Publications (Pub)	378	41	15	631	76	26
Citations (Cit)	2306	307	112	3495	401	155
Cit/Pub	6,10	7,49	7,47	5,54	5,28	5,96

publication which have service focus (i.e., either OI+S, OI+SI or OSI) will results about 5,59 citations in Scopus and 7,02 citations in WoS.

6.2. *Impact based view — Actor level*

The actor level citations analysis is conducted only based on WoS data and presented in Table 5.

As a result about half of the authors do not receive any citations and in all cases having over 10 citations, will results ranking among the top 20% authors. Basically this indicates that only few authors are influential.

The most influential authors in OSI domain are: 1st Chesbrough, Henry (53 citations), 2nd Van Looy, Bart (23), Visnjic Kastalli, Ivanka (23) and (3rd) Bascavusoglu-Moreau, Elif (22), Hughes, Alan (22) and Mina, Andrea (22). Together these are representing 73,3% of citations among all authors.

The most influential authors in OI+SI domain are: 1st Chesbrough, Henry (62), 2nd Mention, Anne-Laure (60) and 3rd Bryson, John R. (56), Love, James H. (56) and Roper, Stephen (56). Together these are representing 46,3% of citations among all authors.

The most influential authors in OI+S domain are: 1st De Jong, Jeroen P. J. (289), De Rochemont, Maurice (289), Van De Vrande,

Table 5. Actor level citation analysis.

Citations	OSI (%)	OI+SI (%)	OI+S (%)	SI (%)	OI (%)
0	42.1	55.2	53.4	55.8	50.2
1 or less	42.1	63.8	64.7	65.8	60.4
2 or less	57.9	71.4	73.6	70.2	66.0
3 or less	68.4	74.3	77.9	74.3	70.5
4 or less	81.6	79.0	79.6	77.0	73.4
5 or less	81.6	79.0	82.2	79.7	75.7
10 or less	81.6	81.0	87.5	85.9	83
50 or less	97.4	95.2	96.7	96.1	95.6

Vareska (289) and Vanhaverbeke, Wim (289), 2nd Lichtenthaler, Ulrich (136), Tajar, Abdelouahid (136) and Tether, Bruce S. (136) and 3rd Dittrich, Koen (135) and Duysters, Geert (135). Together these are representing 27,0% of citations among all authors.

The most influential authors in SI domain are: 1st Ostrom, Amy L. (679), 2nd Bitner, Mary Jo (658) and 3rd Brown, Stephen W. (481). Together these are representing 7,3% of citations among all authors.

The most influential authors in OI domain are: 1st Chesbrough, Henry (2335), 2nd Lichtenthaler, Ulrich (1081) and 3rd Gassmann, Oliver (702). Together these are representing 8,6% of citations among all authors.

6.3. *Discussion*

The impact of OI+S, OI+SI and OSI studies has been steadily growing. This can be considered as a positive indicator for OSI research domain. Over the years the balance in favor for citations in Scopus has been reducing, but as a whole, via Scopus database one can find more influential publications. Thus if only one database will be used for literature reviews, one should favor Scopus over WoS.

Importantly, the studies grounded on OI+SI or OSI have gained only low level citations comparing the studies highlighting services in conjunction with OI term. As a result it is argued that the strategy to highlight OSI term as own entity, might not be the most solid strategy to popularize this research domain. Also the author level analysis indicates that so far only few authors have been influential. The low level citation count combined with only few influential authors verifies that OSI research domain is still in the very early pioneering stage and there is still a lot of room for novel research.

7. Conclusions

In this study, the contemporary body of knowledge of OSI research was described by utilizing various scientometrics methods and comparing the ISI WoS, Scopus and Google Scholar databases. Importantly in this study a clear differences between these data sources was identified. In most cases Scopus had more extensive coverage than WoS, which is in-line with prior studies (Falagas *et al.*, 2008). However, Google Scholar had clearly more extensive coverage which in most cases appeared to have relatively similar profile as WoS and Scopus. As a result it is suggested that innovation management studies should favor Scopus as a data source over WoS when only one citation database is used. Furthermore, Google Scholar was able to detect trends more quickly by identifying faster how OI overcame SI in popularity. Therefore, special attention should be given for Google Scholar when research domain is evolving strongly.

The result suggest that OSI research domain is still infancy and so far there are only handful publications and authors which have embraced the "OSI" term. However, there is clearly a latent potential for OSI studies since publications focusing OI in conjointly with services or SI have been much more popular. Therefore for those who are interested in OSI research it strongly suggested to expand their searches beyond "OSI" term.

Notably, interest towards OI but also SI seems to be a leveling, but there is still strong community of researchers who are involved. Currently OSI as a separate term is still struggling in the pioneering stage and the future forecast for expansion is not looking promising. New authors are not embracing the term, which handful of authors are trying to keep alive and push ahead. In innovation management domain there are already many examples of rivaling terms (Santonen *et al.*, 2016) and therefore it should be carefully considered if there is really a need for separate OSI. The relevant audience can be reach also by incorporating OI with service or SI.

References

Aguillo, I. F. (2012). Is Google Scholar useful for bibliometrics? A webometric analysis. *Scientometrics*, 91(2), 343–351.

Becheikh, N., Landry, R., & Amara, N. (2006). Lessons from innovation empirical studies in the manufacturing sector: A systematic review of the literature from 1993–2003. *Technovation*, 26(5), 644–664.

Chesbrough, H. W. (2006). Open innovation: The new imperative for creating and profiting from technology. Boston, Massachusetts: Harvard Business School Press.

Chesbrough, H. W. (2011). Bringing open innovation to services. *MIT Sloan Management Review*, 52(2), 85.

Choi, J., Yi, S., & Lee, K. C. (2011). Analysis of keyword networks in MIS research and implications for predicting knowledge evolution. *Information & Management*, 48(8), 371–381.

Dahlander, L., & Gann, D. M. (2010). How open is innovation? *Research Policy*, 39(6), 699–709.

De Winter, J. C., Zadpoor, A. A., & Dodou, D. (2014). The expansion of Google Scholar versus Web of Science: A longitudinal study. *Scientometrics*, 98(2), 1547–1565.

Denzin, N. K. (1978). *The Research Act* (2d edn.). McGraw-HI, New York.

Dosi, G. (1982). Technological paradigms and technological trajectories. *Research Policy*, 11(3), 147–162.

Falagas, M. E., Pitsouni, E. I., Malietzis, G. A., & Pappas, G. (2008). Comparison of PubMed, Scopus, web of science, and Google scholar: Strengths and weaknesses. *The FASEB Journal*, 22(2), 338–342.

Gupta, B. M., & Bhattacharya, S. (2004). Bibliometric approach towards mapping the dynamics of science and technology. *DESIDOC Journal of Library & Information Technology*, 24(1), 3–8.

Harzing, A. W., & Van Der Wal, R. (2009). A Google Scholar h-index for journals: An alternative metric to measure journal impact in economics and business. *Journal of the Association for Information Science and Technology*, 60(1), 41–46.

Hood, W., & Wilson, C. (2001). The literature of bibliometrics, scientometrics, and informetrics. *Scientometrics*, 52(2), 291–314.

Hossain, M., & Anees-ur-Rehman, M. (2016). Open innovation: An analysis of twelve years of research. *Strategic Outsourcing: An International Journal*, 9(1), 22–37.

Hossain, M., Islam, K. Z., Sayeed, M. A., & Kauranen, I. (2016). A comprehensive review of open innovation literature. *Journal of Science & Technology Policy Management*, 7(1), 2–25.

Huizingh, E. K. (2011) Open innovation: State of the art and future perspectives. *Technovation*, 31(1), 2–9.

Kovacs, A., Van Looy, B., & Cassiman, B. (2015). Exploring the scope of open innovation: A bibliometric review of a decade of research. *Scientometrics*, 104(3), 951–983.

Kuhn, T. S. (1962). *The Structure of Scientific Revolutions*. University of Chicago Press, Chicago, IL.

Larivière, V., Sugimoto, C. R., & Cronin, B. (2012). A bibliometric chronicling of library and information science's first hundred years. *Journal of the American Society for Information Science and Technology*, 63(5), 997–1016.

Lisée, C., Larivière, V., & Archambault, É. (2008). Conference proceedings as a source of scientific information: A bibliometric analysis. *Journal of the American Society for Information Science and Technology*, 59(11), 1776–1784.

McLean, L. D. (2005). Organizational culture's influence on creativity and innovation: A review of the literature and implications for human resource development. *Advances in Developing Human Resources*, 7(2), 226–246.

Medeiros, G., Binotto, E., Caleman, S., & Florindo, T. (2016). Open innovation in agrifood chain: A systematic review. *Journal of Technology Management & Innovation*, 11(3), 108–116.

Meho, L. I., & Yang, K. (2007). Impact of data sources on citation counts and rankings of LIS faculty: Web of Science versus Scopus and Google Scholar. *Journal of the Association for Information Science and Technology*, 58(13), 2105–2125.

Mingers, J., & Lipitakis, E. A. (2010). Counting the citations: A comparison of Web of Science and Google Scholar in the field of business and management. *Scientometrics*, 85(2), 613–625.

Moore, J. F. (1996). *The Death of Competition: Leadership & Strategy in the Age of Business Ecosystems*.HarperBusiness, New York.

Nacke, O. (1979). *Informetrie: Ein neuer Name für eine neue Disziplin, Nachrichten für Dokumentation*, 30, 212–226.

Pilkington, A., & Meredith, J. (2009). The evolution of the intellectual structure of operations management — 1980–2006: A citation/co-citation analysis. *Journal of Operations Management*, 27(3), 185–202.

Pritchard, A. (1969). Statistical bibliography or bibliometrics? *Journal of Documentation*, 25, 348–349.

Randhawa, K., Wilden, R., & Hohberger, J. (2016). A bibliometric review of open innovation: Setting a research agenda. *Journal of Product Innovation Management*, 33(6), 750–772.

Santonen, T., & Conn, S. (2015a). Evolution of research topic at ISPIM from 2009 to 2014. In *Proceedings of The ISPIM Innovation Summit*, 6–9 December, Brisbane, Australia.

Santonen, T., & Conn, S., (2015b). Research topics at ISPIM: Popularity-based Scientometrics keyword analysis. In H. Eelko, T. Marko, C. Steffen, B. Iain (eds.), *The Proceedings of the XXVI ISPIM Innovation Conference*, 14–17 June, Budapest, Hungary.

Santonen, T., & Conn, S., (2016). Social network analysis based keyword analysis of ISPIM research topics. In: *The XXVII ISPIM Innovation Conference — Blending Tomorrow's Innovation Vintage*, 19–22 June, Porto, Portugal. LUT Scientific and Expertise Publications, Reports.

Santonen, T., Normann Kristiansen J., & Gertsen, F. (2016). Increased variation or higher fences? understanding typological evolution in radical innovation management. In: *The XXVII ISPIM Innovation Conference — Blending Tomorrow's Innovation Vintage*, 19–22 June, Porto, Portugal.

Santonen, T. (2018). Comparing living lab(s) and its' competing terms popularity in ISPIM Publications. In I. Bitran, S. Conn, K.R.E. Huizingh, O. Kokshagina, M. Torkkeli, and M. Tynnhammar (eds.), *Proceedings of the 2018 ISPIM Innovation Conference (Stockholm): Innovation, the Name of the Game in Stockholm*, 17–20 June, Sweden. LUT Scientific and Expertise Publications, Reports.

Santonen, T., Tynnhammar, M., & Conn, S. (2018). A review of research methods in ISPIM publications. In: I. Bitran, S. Conn, K.R.E. Huizingh, O. Kokshagina, M. Torkkeli, and M. Tynnhammar (eds.), *Proceedings of the 2018 ISPIM Innovation Conference (Stockholm): Innovation, the Name of the Game in Stockholm*, 17–20 June, Sweden. LUT Scientific and Expertise Publications, Reports.

Smith, H. W. (1975). *Strategies of Social Research: The Methodological Imagination.* Prentice Hall, Englewood Cliffs, NJ.

Su, H. N., & Lee, P. C. (2012). Framing the structure of global open innovation research. *Journal of Informetrics*, 6(2), 202–216.

Van Raan, A. F. J. (Ed.) (1998). Special topic issue: Science and technology indicators. *Journal of the American Society for Information Science*, 49, 3–81.

Vilko, J., & Santonen, T. (2017). Innovation and risk management — A cute couple or opposing forces? In: I. Bitran, S. Conn, K.R.E. Huizingh, O. Kokshagina, M. Torkkeli, and M. Tynnhammar (eds.), *Proceedings of the ISPIM Innovation Summit 2017 (Melbourne): "Building the Innovation Century"*, 10–13 December, RMIT University, Melbourne, Australia. LUT Scientific and Expertise Publications, Reports.

Wasserman, S., & Faust, K. (1994). *Social Network Analysis: Methods and Applications.* Cambridge University Press, Cambridge.

Yi, S., & Choi, J. (2012). The organization of scientific knowledge: The structural characteristics of keyword networks. *Scientometrics*, 90(3), 1015–1026.

Zhao, Y., & Zhu, Q. (2012). Evaluation on crowdsourcing research: Current status and future direction. *Information Systems Frontiers*, Published online: 11 April 2012.

https://doi.org/10.1142/9789811234491_0004

Chapter 4

Business Models for Collaborative eHealth in Homecare*

Niels F. Garmann-Johnsen[†,§] *and Santiago Martinez*[‡,¶]

[†]*University of Agder, Gimlemoen 19, 4630 Kristiansand, Norway*

[‡]*University of Agder, Campus Kristiansand, Universitetsveien 25, 4630 Kristiansand, Norway*

[§]*niels.f.garmann-johnsen@uia.no*

[¶]*santiago.martinez@uia.no*

Abstract. eHealth innovations promise to address the many challenges to the welfare system posed by demographic changes in the population, by connecting services from different organizations through collaborative online networks. Open innovation and co-creating with partners and healthcare users may boost such transformations, but there is has been a gap in literature on how Open innovation can be managed in a public sector healthcare setting; how it's principles can help overcoming inertia, achieving

*This chapter builds in part on: N. F. Garmann-Johnsen and S. Martinez (2017). Roadmap for collaborative eHealth service architectures for homecare — general ebusiness requirements. In: The Proceedings of: eTELEMED 2017, The Ninth International Conference on eHealth, Telemedicine, and Social Medicine (ISBN: 978-1-61208-540-1).

functional fit and creating new values for healthcare users and benefits for the organization. Looking at literature on how network business models has been applied to healthcare, the authors elicit propositions for new eHealth solutions design, incorporation Open innovation principles.

Keywords. eHealth; homecare; collaboration; co-innovation.

1. Introduction

In many industrialized countries, such as Norway, one of the main goals for eHealth innovation is the use of information and communication technologies (ICT) and information systems (IS) to avoid the inflection point in healthcare: it is estimated that by the year 2025, the demand of care could not be sufficiently covered based on today's healthcare delivery (Gaba, 2004). Change and innovation are needed for treatments maintaining the quality of care within assigned budgets, where the resources needed transcend those of one singular organization. Using technology that helps people with care needs to stay safely, longer at home and out of institutions and hospitals may be a strategy to reduce the total healthcare costs of society (Garmann-Johnsen, 2015). There has been a lack of open innovation research in public contexts, and it is necessary to perform further research with a focus on how open innovation can be managed in healthcare (Wass & Vimarlund, 2016). This book will contribute to closing this gap, where this chapter looks deeper at the case of eHealth in homecare.

The paradigm and capability of open innovation (Chesbrough & Bogers, 2014) where ideas and information flow in managed flows in and out of the focused service-organization pursues building "positional assets" and other "dynamic capabilities" (Teece *et al.*, 1997) through collaboration (Teece *et al.*, 1997; Garmann-Johnsen *et al.*, 2017). However, knowledge of methods and roadmaps are needed for the organization to proceed to build on such affordances.

The authors present a roadmap for service design and architecture development in collaboration across organizations in healthcare. The research focuses in the home care, leaving clinical care outside the scope of this study. A challenge associated to the home care is that it presents a broad spectrum of user requirements within the same information technology (IT) infrastructure. Research in eBusiness models may inform future strategies and research in eHealth service design, especially in homecare. For the matter of this chapter, the authors reviewed the literature related to (e-)business models and eHealth for later on offering propositions about service design.

Collaboration and open innovation are closely linked. Open innovation entail opening the service-organization for new ideas for products and services. Open innovation processes are often illustrated as a funnel leading from research to marketing, but narrowing down the field of selected ideas or concepts as one focuses towards an application. The study of stakeholder requirements and successful business model principles could guide such a reduction. Gould (2012) offers a model for combining stakeholder involvement in the process of open innovation. Such a process requires dialog and sharing information with outside stakeholders as well as collecting information. The process can thus create both foreseen and unforeseen values for users and stakeholders.

Studies into stakeholder requirements in eHealth innovation and pre-procurement (Mettler & Eurich, 2012) points to the need for more stakeholder involvement. Furthermore, design principles for business models that can facilitate such involvement are here elicited. The need in homecare is to develop an adaptive framework that allows front-line personnel (e.g., nurse, general practitioner) and other end-users to dynamically manage and (re-)configure business processes, patient services and health treatment clinical pathways, with the purpose of creating service-oriented solutions whose patient autonomy and interactivity are core elements.

The main research questions targeted in this chapter is to find in the research literature general business requirements that influence eHealth services in home care. This leads to the two research questions addressed in this paper:

RQ1. What are the business requirements for enterprise systems (ESs) that create value for users in the context of open innovation (Chesbrough & Rosenbloom, 2002) in eHealth?

RQ2. What are the preconditions to gain benefits from new ESs and increase efficiency and quality in eHealth adoption?

This chapter is structured in five sections. Section 2 presents the theoretical framework that supports the content of this chapter. Section 3 described the method employed. Section 4 presents the results including the answers to the research questions presented above. Section 5 draws the conclusion and future work.

2. Theoretical Framework

Teece *et al.* (1997) tended to see an organization's business model as a rarely changed positional asset. Today, many researchers tend to see the development of an organizations business model(s) as a dynamic process (Garmann-Johnsen & Hellang, 2014). Moreover, an organization in a so-called "n-sided" market with multiple services may have more than one business model (Porter, 1991). A successful business model unlocks the value creating potential of processes and technologies (Chesbrough & Rosenbloom, 2002). Following the principles of open innovation of import of ideas, an organization may succeed with a business model where others have failed. This poses a relevant question, what are the business models candidate to succeed within homecare services, and which elements of those business model can be favorably adopted in general terms?

There is a variety of business models, but Hwang (2009) proposes three main types or business model templates that help to categorize the business processes of modern enterprises better than the traditional value chain model of the firm (Stabell & Fjeldstad, 1998) alone:

1. *Solution shop*. A place where customers describe their problem and it can be fixed. For example, general practitioners and hospitals curing patients.
2. *Value-adding process* (comparable to the value chain) typical of manufacturing enterprises; comparable to Porters' (Porter, 1991) model of the firm. For example, pharmaceutical producers.
3. *Facilitated user networks* (value networks) to connect users and services. For example, different web communities (most communication- and internet-based businesses fall under this category).

With regard to these models, Hwang (2009), building on works by Stabell and Fjeldstad (1998), states that the facilitated user network is under-represented in the health sector — and, thus, represents growth potential: as more businesses move online, facilitated user networks represent a business model template of increasing significance, including eHealth. Thus the authors will explore these concepts further in relation to eHealth service architecture in an environment of inter-organizational collaboration and co-creation. Frameworks for analyzing value creation in e-Business (e.g., Amit and Zott, 2001) are relevant for understanding the eHealth landscape because many eHealth systems contain elements of e-Business models which belong to the category of Value Networks.

Amit and Zott's (2001) e-business value creation model categorizes the sources of value creation into four main groups: *efficiency*, *novelty*, *lock-in* and *complementarity*. These groups are relevant to e-business models that often implement instances of the facilitated user network archetype of business models, and inter-organizational process collaborations. According to Amit and Zott, *efficiency* has been one of the primary driving forces for the transition to a more digital economy in all phases of the transaction. *Complementarities* that arrive from the bundle of goods provide more value to the users than the sum of values provided by each good alone. In business-to-business relationships, complementarity signifies collaborative networks, "enriching the company's own knowledge base through the integration of suppliers, customers, and external knowledge sourcing" (Enkel *et al.*, 2009). Complementarity and novelty imply collaboration and innovation — both themes of significance in healthcare. *Novelty* means that there is a "first" here, either in the service itself or in the way that it is conceived or delivered. Combining or connecting two or more previously unconnected parties, domains or methods often achieves this novelty. A *lock-in* value is created through users' reuse of one-time transactions costs (e.g., for registration and service customization).

Seddon *et al.* (2010) proposed a model that explains how organizations, such as municipalities, can overcome organizational inertia and successfully adopt new ESs and obtain benefits, e.g., quality and efficiency. ESs support processes, information flows, reporting and business analytics within and between organizations and individuals. Specifically, these processes support organizations' business models. Business models are models of how enterprises create value for themselves and other, while also capturing value. An organization can have more than one business model, and business models can include more than one organization.

Since RQ2 aims to explore the affordance of ES in realizing emergent business models in eHealth, the authors have chosen to use Seddon *et al.*'s model of key factors affecting organizational benefits from ES (OBES) (Seddon *et al.*, 2010):

1. Short-term/on-going business improvement projects:
   a. Functional fit (FF)
   b. Overcoming organizational inertia (OOI)
2. With the addition of long-term organizational benefit factors:
   a. Integration
   b. Process optimization
   c. Improved access to information

Functional fit of software means that the processes supported by the ES are efficient and effective for the organization and that they help people to finish their tasks. Overcoming organizational inertia is the extent to which organizations' members are motivated to learn, use and accept new systems. Integration refers to the unification of systems and/or data from resources within and outside an organization. Tailoring the IS to an organization's goals help to achieve process optimization. This often necessitates improved access to information.

3. Method

The authors made a systematic literature review about business models in eHealth using Amit and Zott's (2001) framework for value creation in eBusiness to answer the first research question (RQ1), and conceptualize the requirements. Then, we analyze concepts found in the review with the model proposed by Seddon *et al.* (2010) to answer the second research question (RQ2), and produce general propositions for information system design with special regards to primary care and homecare.

The literature search was performed in spring 2014, using the search engines ProQuest and EBSCOhost. The search engines targeted science and social science disciplines. The authors used truncations *health* AND (*ecommerce* OR *reward*) AND "business models" in the search terms. The search included scholarly papers and trade journals, choosing to include the latter due to the "newness" of the area in explored. The authors found 286 papers that matched the search criteria. However, a screening showed that 246 of the found papers were not related with eHealth business models. These papers were excluded. The excluded papers were concerned with topics such as: clinical trials, environmental issues, other non-health-related issues and management issues at large.

The remaining 40 papers (34 found through ProQuest and six found through EBSCOhost) were rated as relevant to the research topic and, thus, analyzed. A table showing the results and list of the searched and reviewed articles is included in the appendix to this chapter.

One of the more interesting concepts turning up in our initial research was "co-innovation", that it is explained in the next section. We updated the search with articles from 2014 up to July 2017, using the search terms "co-innovation", "eHealth" and "business models". This search returned 39 results of which five articles contributed with new insights, of relevance to our research questions, cited in the introduction and/or in the results section.

4. Results

4.1. *Answer to RQ1*

This section shows the findings from the systematic literature review, which were aggregated and conceptualized following the guidance of Webster and Watson (2002), and then sorted according to Amit and Zott's model. The authors added the category "Other" to collect concepts that were not easily sorted under Amit and Zott's four main sources; these were predominantly societal and public policy factors. These other sources stemmed from the special public or semi-public financial context of eHealth in countries such as United Kingdom and Scandinavian ones.

(1) **Efficiency concepts**

- *Easier scaling and sharing*: New eHealth concepts leverage the efficiency and cost-flexibility of cloud computing. Cloud computing facilitates the use of intelligent algorithms to detect heightened risks of worsening conditions and supports the triage classification of patients (triage is a priority assessment of health treatment for patients). Such capabilities can be shared and utilized by several organizations. Standardizations and reductions in the number of product lines greatly reduce overhead cost ratios. Companies must defend low costs against overly personalized service.
- *Transaction cost efficiency*: Through e-business' transaction efficiency, doctors can now charge costs associated to phone calls and online consultation. Cloud computing also creates the opportunity for self-service solutions. Online solutions also enable easier care and quality coordination. *Easier sourcing*: Cloud services provide easier possibilities for sourcing support functions (e.g., system running).

(2) **Novel concepts**

- *Rapid customization and co-innovation*: The cloud model allows for rapid prototyping and innovation in processes, products and services. Co-innovation and co-creation generate new organizational and shared values. Surplus innovation can be sold-off, generating rent (commercial profit). Moreover, the possibility of doctor "visits" via secure video-conferencing opens for new possibilities for patients and health professionals. Partnerships with educational providers provide focused and personalized professional training.

(3) **Lock-In concepts**

- *Revenue and reward splits*: Cloud services can store and make use of user preferences, which can enable product and service customization. Offering a choice of market channels to consumers can be mutually beneficial for all parties, due to the aggregation of products and services. For instance, the Hello Health website (Kravitz, 2013) negotiates a revenue split of subscriptions with physicians. Sharing information across organizational borders creates better possibilities for personalized health coaching.
- *Fun and belonging*: Online solutions can leverage the network and "gamification" (i.e., comparing progress with peers) effects of social media. Algorithms for the content and sentiment analysis of entries in online communities can suggest alternatives or interest groups that might be beneficial to join. Corporations should link medical plans for employees with corporate strategies, which could, for example, reward the use of fitness facilities and programs.

(4) **Complementarity concepts**

- *Aggregating services*: Internet offers possibilities for aggregating and sourcing services and finance. For instance, a combination of many travel-related and medical businesses and services can facilitate medical tourism.
- *Shared development risk*: Alliance management capabilities have positive effects on research and development

capabilities. For example, in targeted medicine, there is a call for cross-business collaboration. Targeted medicine makes it possible to treat rare, previously untreatable conditions; however, the cost per patient is high, and some governments may refuse reimbursement. By reducing development time through improved collaborations between test laboratories and pharmaceutical companies, targeted medicine can be achieved at manageable costs. Pharmaceutical companies must learn how to fit into other companies' business models. All actors need to be satisfied, for example with "win–win" scenarios.

- *Management coordination*: Primary and secondary care must be better coordinated. Companies' external and internal resources must be aligned in order to facilitate the internalization of knowledge. Many user networks' business models are without profit or low profit; however, as a potential source of empowerment of patients, they represent ideal opportunities for business models for treating chronic diseases. These are costly treatments and are not well suited for the traditional business model of hospitals and physicians.

(5) **Other effects: Public and societal concepts**

- *New reimbursement schemes for better coordination*: Payment models in health should encourage coordination. Reimbursement models should encourage preventive measures, such as Healthy Outlook (Micheli *et al.*, 2012). They should also reward the results of coherent treatment over "pay-per-consultation" systems. New reimbursement schemes may pose new challenges for financial reporting.
- *Knowledge management*: Knowledge benefits from exchange, rather than protection. The best "players" build "invisible" information infrastructures for developing, for example, cancer-focused IT system and evidence-based medicines.
- *Ethical compliance*: Ethical compliance management is of growing significance in a changing eHealth business environment. Better management can be achieved through collaboration and democratic governance.

4.2. *Answer to RQ2*

Our extended review provided insight into what preconditions are necessary to achieve benefits from new (business and healthcare) systems and succeed with overcoming inertia. These propositions refer to external factors associated to co-innovation/open innovation, and internal factors such as management-systems. These factors are, however, arguably closely linked.

Adequate business models and service designs that generate successful partnerships is a precondition for success. But the question is, how do we get there? Switching from a pure techno-centric approach to healthcare to a socio-technical approach, giving equal attention to both social and technical aspects of complex systems, will benefit design and implementation of new services (Beaumont *et al.*, 2014). The way in which new technological artifacts are commercialized is the deciding factor for the economic outcome (Volland & Eurich, 2014). Sustainable competitive advantages in business, or sustainable financing of public services, require the continual delivery of maximum value to the individual customer or user (García Haro *et al.*, 2014). One needs to explore, experiment, evaluate and co-create with user-communities (Mercier-Laurent, 2015), where an example of such technique can be found in the Living Lab methodology.

Based on the business concepts induced from literature and rendered in the previous section, the authors thus propose the following propositions as design principles for a Municipal eHealth architecture:

Proposition 1 (Shared e-commerce solutions). *E-commerce solutions shared among collaborating partners, such as Hello Health* (Kravitz, 2013), *enable easier scaling and sharing, greater transaction cost efficiency and easier sourcing of non-strategic support functions, such as system running (i.e., application-server management). Shared e-commerce solutions support short-term functional fit and long-term business benefits, providing integration, process optimization and improved access to information.*

Proposition 2 (Co-creation with partners and clients (and patients)). *By providing customers and partners alternatives for customizing their services and by encouraging and proactively collecting feedback on services, it is possible to distinguish patterns that provide an impetus for improving services and creating new ones. Co-creation may produce positive short- and long-term effects.*

Proposition 3 (Aligning internal and external resources). *To overcome organizational inertia, structural changes may become necessary. Internal organizational resources should be dedicated to match external resources (business partners) in order to extract and embed external knowledge within an organization. This process is connected to Propositions 4, 5 and 7.*

Proposition 4 (Aligning price and reward (bonuses and revenue sharing)). *Aligning rewards and bonus schemes through the delivery chain and throughout user levels encourages the achievement of goals for treatment and health policies.*

Proposition 5 (Iterations for the economic alignment of interests). *Throughout a delivery chain, there is information asymmetry with regard to knowledge of processes and costs. Over time, all actors in the chain experience a learning curve; thus, collaborative iterations are needed to optimize business processes and to maintain win–win situations throughout the delivery chain. The role of ES is to provide accurate, reliable and comparable data for collaborative decision-making, thereby stimulating and supporting organizational learning.*

Proposition 6 (Social media integration and capabilities). *To ensure that users and patients choose digital solutions and achieve the efficiency benefits that such solutions provide, digital solutions should ideally be socially oriented and easy and fun to use. Gamification allows users to compare their personal data in fitness or treatment programs — and, thus, encourages and motivates users/patients to adapt and use solutions. Gamification can be developed and* *stimulated through social networks. Actors in healthcare (e.g., hospitals, primary care) need to develop their social media strategies* (Effing and Spil, 2016).

Proposition 7 (Alignment of incentives). *A strong reallocation of Intellectual Property Rights and revenue sources may be necessary to encourage more efficient research and development and to achieve joint goals for development, thereby lowering the costs of new treatments (e.g., targeted medicine). Increased use of public–private collaboration contracts around research and development might be potentially considered.*

Proposition 8 (Transparency (sharing information across different levels)). *Often, a patient is not the one who pays the full price for a medical service. In many economies, the state or company medical plans are the ones who pay for such services via voluntary or compulsory insurance. Demographic changes, combined with improving treatment options, have led to rising costs of healthcare. To improve this situation, reimbursement schemes increasingly seek possibilities to pay for results rather than consultations. For health providers, this calls for increased information sharing, and transparency of treatment results. This proposition is connected to Proposition 9.*

Proposition 9 (Associating performance measurement and finance). *Enterprise/planning systems of medical providers and vendors must increasingly be able to provide data for quality and financial performance to both market and government. Business intelligence and balanced scorecard systems are needed for quality and performance management, and they can be directly linked to reimbursement schemes in the future.*

Proposition 10 (Democratic governance). *Ethical compliance is gaining importance in many areas, and it has great significance in eHealth* (Rossi, 2010). *A proactive strategy for compliance management, involving consistent policies for handling legal, quality and treatment deviations or complaints, including privacy and security*

Table 1. Search strategy, reviewed articles.

Search Strategy						
Date	**Database**	**Search string**	**In**	**Constraints**	**F1**	**F2**
20/05/14	ABI/Inform (ProQuest), all databases	*health* AND (*ecommerce* OR *reward*) AND "business models"	Scholarly journals (244), trade journals (3)	2008–2014, peer-reviewed	247	34
20/05/14	EBSCOhost all databases	*health* AND (*commerce* OR *reward*) AND "business models"	Scholarly- and trade journals	2008–2014	39	6
SUM					**286**	**40**

Note: F1 = Total number of articles, F2 = Articles containing relevant knowledge for this study.

List reviewed articles (F2): Andersson *et al.* (2009), Badinelli *et al.* (2012), Berman *et al.* (2012), Berryman *et al.* (2013), Britnell (2011), Cai and Chen, (2011), Case (2011), Chernew *et al.* (2011), Christiansen *et al.* (2013), Coye *et al.* (2009), Den Hertog *et al.* (2010), DeVore and Champion (2011), Dilwali (2013), Edlin (2013), Ernst&YoungCorp. (2010), Garrison (2009), Goroff and Reich (2010), Granger *et al.* (2012), Kasabov and Warlow (2010), Koelsch *et al.* (2013), Koster (2013), Kravitz (2013), Lee *et al.* (2012), Lester (2009), Lukkari (2011), Main (2011), Majette (2009), Micheli *et al.* (2012), Nielsen (2007), Prior (2008), Purdy *et al.* (2012), Rossi (2010), Shah (2012), Smith *et al.* (2011), Sorrentino and Garraffo (2012), Swan (2012), Tang and Lee (2009), Tencati and Zsolnai (2009), Verhoef *et al.* (2009), Zismer (2013).

issues, may benefit an organization in the long term. This may help overcoming barriers to e-Health adoption. Actively involving users and employees in this work through democratic governance can enable such a result (Rossi, 2010) (Table 1).

5. Conclusions

This study has answered the two research questions (RQs) by eliciting business models and ESs design principles that may inform future eHealth architectures and applications. Although these principles are not exhaustive, they illustrate the need for flexibility and capability of co-creation with users at all levels.

The found propositions can be tested in a Living Lab test setting (Pallot *et al.*, 2010), with special regard to adaptability, flexibility and facility for user-co-creation in design of eHealth services. With this study we aimed to contribute to the open innovation literature, with practical methods and propositions for successful implementation of new services in homecare and similar public services. Based on the empirical findings from the case-studies found in the literature, a generalization of a pre-procurement design process-model is made, consisting of a roadmap of co-creation processes that combine the goals of all stakeholders; application-vendors, municipalities, their constituents and end-users. The exploration of the connection between innovations in local media and applications and central information infrastructures (Software as a Service, cloud computing solutions) is suggested as a future work consequent with the results presented in this chapter.

References

Amit, R., & Zott, C. (2001). Value creation in e-business. *Strategic Management Journal*, 22(6–7), 493–520.

Andersson, B., Johannesson, P., & Zdravkovic, J. (2009). Aligning goals and services through goal and business modelling. *Information Systems and e-Business Management,* 7(2), 143–169.

Badinelli, R., Barile, S., Ng, I., Polese, F., Saviano, M., & Di Nauta, P. (2012). Viable service systems and decision making in service management. *Journal of Service Management,* 23(4), 498–526.

Beaumont, L. C., Bolton, L. E., McKay, A., & Hughes, H. P. N. (2014). Rethinking service design: A socio-technical approach to the development of business models. In: *Product Development in the Socio-sphere.* Springer International Publishing, pp. 121–141.

Berman, S. J., Kesterson-Townes, L., Marshall, A., & Srivathsa, R. (2012). How cloud computing enables process and business model innovation. *Strategy & Leadership,* 40(4), 27–35.

Berryman, S. N., Palmer, S. P., Kohl, J. E., & Parham, J. S. (2013). Medical home model of patient-centered health care. *Medsurg nursing: Official Journal of the Academy of Medical-Surgical Nurses,* 22(3), 166.

Britnell, M. (2011). The role of the "specialist" in healthcare. *Clinical Medicine,* 11(4), 329–331.

Cai, G. G., & Chen, Y.-J. (2011). In-store referrals on the internet. *Journal of Retailing,* 87(4), 563–578.

Case, I. (2011). A disruptive approach: From education to rewards, UMR's Bart Halling connects benefits blueprints to corporate strategies. *Benefits Selling.* Retrieved May, 2014, from http://www.thefreelibrary.com/A+disruptive+approach%3a+from+education+to+rewards%2c+UMR's+Bart+Halling...-a0347404047

Chernew, M. E., Mechanic, R. E., Landon, B. E., & Safran, D. G. (2011). Private-payer innovation in Massachusetts: The "alternative quality contract". *Health Affairs,* 30(1), 51–61.

Chesbrough, H., & Bogers, M. (2014). Explicating open innovation: Clarifying an emerging paradigm for understanding innovation. In: H. Chesbrough, W. Vanhaverbeke, and J. West (eds.), *New Frontiers in Open Innovation,* Oxford University Press, Oxford, pp. 3–28.

Chesbrough, H., & Rosenbloom, R. S. (2002). The role of the business model in capturing value from innovation: Evidence from Xerox Corporation's technology spin-off companies. *Industrial and Corporate Change,* 11(3), 529–555.

Christiansen, J. K., Gasparin, M., & Varnes, C. J. (2013). Improving design with open innovation: A flexible management technology. *Research-Technology Management,* 56(2), 36–44.

Coye, M. J., Haselkorn, A., & DeMello, S. (2009). Remote patient management: Technology-enabled innovation and evolving business models for chronic disease care. *Health Affairs,* 28(1), 126–135.

Den Hertog, P., Van der Aa, W., & de Jong, M. W. (2010). Capabilities for managing service innovation: Towards a conceptual framework. *Journal of Service Management,* 21(4), 490–514.

DeVore, S., & Champion, R. W. (2011). Driving population health through accountable care organizations. *Health Affairs,* 30(1), 41–50.

Dilwali, P. K. (2013). *From Acute Care to Home Care: The Evolution of Hospital Responsibility and Rationale for Increased Vertical Integration.* Retrieved May 2014, from http://www.ncbi.nlm.nih.gov/pubmed/24396947

Edlin, M. (2013). *State of the Industry: Future of ACOs.* from http://managedhealthcareexecutive.modernmedicine.com/managed-healthcare-executive/news/state-industry-future-acos

Effing, R., & Spil, T. A. (2016). The social strategy cone: Towards a framework for evaluating social media strategies. *International Journal of Information Management,* 36(1), 1–8.

Enkel, E., Gassmann, O., & Chesbrough, H. (2009). Open R&D and open innovation: Exploring the phenomenon. *R&d Management,* 39(4), 311–316.

Ernst&YoungCorp. (2010). Are you ready for Pharma 3.0? August 1, 2010 *Pharmaceutical Technology Europe*, 22(8). Retrieved Feb. 2021, from https://pdf4pro.com/view/global-pharmaceutical-industry-report-ey-4a87a4.html.

Gaba, D. M. (2004). The future vision of simulation in health care. *Quality and Safety in Health Care*, 13(1), i2–i10.

García Haro, M. Á., Martínez Ruiz, M. P., & Martínez Cañas, R. (2014). The effects of the value co-creation process on the consumer and the company. *Expert Journal of Marketing*, 2(2), 68–81.

Garmann-Johnsen, N. F. (2015). What seems to be the problem? — A study of connections between national contexts and regional e-health strategies. *Health Policy and Technology*, 4(2), 144–155.

Garmann-Johnsen, N. F., & Hellang, Ø. (Eds.)(2014). Collaborative process modelling and evaluation in e-health. Proceedings from Scandinavian Conference on Health Informatics, August 21–22, 2014, Grimstad, Norway" (Available at uia.brage.unit.no, accessed Jan. 2021).

Garmann-Johnsen, N. F., & Tom, R. E. (2017). Dynamic capabilities in e-health innovation: Implications for policies. *Health Policy and Technology* (*Journal*) 6(3): 292–301.

Garrison, L. P. (2009). Will pharmacogenomics disrupt the US health care system? No. *Public Health Genomics,* 12(3), 185–190.

Goroff, M., & Reich, M. R. (2010). Partnerships to provide care and medicine for chronic diseases: A model for emerging markets. *Health Affairs,* 29(12), 2206–2213.

Granger, B. B., Prvu-Bettger, J., Aucoin, J., Fuchs, M. A., Mitchell, P. H., Holditch-Davis, D., ... Gilliss, C. L. (2012). An Academic-Health Service Partnership in Nursing: Lessons From the Field. *Journal of Nursing Scholarship,* 44(1), 71–79.

Hwang, J. (2009). Keynote address-the innovator's prescription: An examination of the future of health care through the lenses of disruptive innovation. *Archives of Pathology & Laboratory Medicine,* 133(4), 513–520.

Hwang, J. (2009). The innovator's prescription: An examination of the future of health care through the lenses of disruptive innovation. *Archives of Pathology & Laboratory Medicine,* 133(4), 513–520.

Kasabov, E., & Warlow, A. J. (2010). Towards a new model of "customer compliance" service provision. *European Journal of Marketing*, 44(6), 700–729.

Koelsch, C., Przewrocka, J., & Keeling, P. (2013). Towards a balanced value business model for personalized medicine: An outlook. *Pharmacogenomics*, 14(1), 89–102.

Koster, K. (2013). Incentives: Putting the "cent" in incentives; employers, experts pay closer attention to gaining best ROI on expenditures for wellness incentives. *Highbeam Research.* Retrieved May 2014, from http://www.highbeam.com/doc/1G1-328244567.html.

Kravitz, R. L. (2013). Improvement happens: A commercial IT solution for reviving primary care? An interview with Hello Health CEO Nathanial Findlay and colleagues. *Journal of General Internal Medicine*, 28(2), 310–314.

Lee, S. M., Olson, D. L., & Trimi, S. (2012). Co-innovation: Convergenomics, collaboration, and co-creation for organizational values. *Management Decision*, 50(5), 817–831.

Lester, D. S. (2009). Will personalized medicine help in "transforming" the business of healthcare? *Personalized Medicine*, 6(5), 555–565.

Lukkari, P. (2011). Merger: Institutional interplay with customer relationship management. *Management Research Review*, 34(1), 17–33.

Main, T. (2011). Cancer and healthcare reform: Making the pieces fit. *Oncology Journal*, 783.

Majette, G. R. (2009). *From Concierge Medicine to Patient-Centered Medical Homes: International Lessons & The Search for A Better Way to Deliver Primary Health Care in the U.S.* Retrieved May 2014, from http://www.ncbi.nlm.nih.gov/pubmed/20196284.

Mercier-Laurent, E. (2015). *The Innovation Biosphere: Planet and Brains in the Digital Era*. John Wiley & Sons.

Mettler, T., & Eurich, M. (2012). A "design-pattern"-based approach for analyzing e-health business models. *Health Policy and Technology*, 1(2), 77–85.

Micheli, P., Schoeman, M., Baxter, D., & Goffin, K. (2012). New business models for public-sector innovation: Successful technological innovation for government. *Research-Technology Management*, 55(5), 51–57.

Micheli, P., Schoeman, M., Baxter, D., & Goffin, K. (2012). New business models for public-sector innovation: Successful technological innovation for government. *Research-Technology Management*, 55(5), 51–57.

Nielsen, C. (2007). A content analysis of analyst research: Health care through the eyes of analysts. *Journal of Health Care Finance*, 34(3), 66–90.

Pallot, M., Trousse, B., Senach, B., & Scapin, D. (Eds.) (2010). Living lab research landscape: From user centred design and user experience towards user cocreation. *First European Summer School Living Labs*.

Porter, M. E. (1991). Towards a dynamic theory of strategy. *Strategic Management Journal*, 12(2), 95–117.

Prior, L. (2008). Repositioning documents in social research. *Sociology*, 42(5), 821–836.

Purdy, M., Robinson, M. C., & Wei, K. (2012). Three new business models for "the open firm". *Strategy & Leadership*, 40(6), 36–41.

Rossi, C. L. (2010). Compliance: An over-looked business strategy. *International Journal of Social Economics*, 37(10), 816–831.

Rossi, C. L. (2010). Compliance: An over-looked business strategy. *International Journal of Social Economics*, 37(10), 816–831.

Seddon, P. B., Calvert, C., & Yang, S. (2010). A multi-project model of key factors affecting organizational benefits from enterprise systems. *MIS Quarterly*, 34(2), 305–328.

Shah, N. (2012). *E-Commerce and ePayment, There is Still a Lot of Untapped Potential*. Retrieved May 2014, from http://www.scribd.com/doc/97771063/Siliconindia-June-12-Issue.

Smith, P., Hampson, L., Scott, J., & Bower, K. (2011). Introducing innovation in a management development programme for a UK primary care organisation. *Journal of Health Organization and Management*, 25(3), 261–280.

Sorrentino, F., & Garraffo, F. (2012). Explaining performing R&D through alliances: Implications for the business model of Italian dedicated biotech firms. *Journal of Management & Governance*, 16(3), 449–475.

Stabell, C. B., & Fjeldstad, Ø. D. (1998). Configuring value for competitive advantage: On chains, shops, and networks. *Strategic Management Journal*, 19(5), 413–437.

Swan, M. (2012). Scaling crowdsourced health studies: The emergence of a new form of contract research organization. *Personalized Medicine*, 9(2), 223–234.

Tang, P. C., & Lee, T. H. (2009). Your doctor's office or the Internet? Two paths to personal health records. *New England Journal of Medicine*, 360(13), 1276–1278.

Teece, D. J., Pisano, G., & Shuen, A. (1997). Dynamic capabilities and strategic management. *Strategic Management Journal*, 18(7), 509–533.

Tencati, A., & Zsolnai, L. (2009). The collaborative enterprise. *Journal of Business Ethics*, 85(3), 367–376.

Verhoef, P. C., Lemon, K. N., Parasuraman, A., Roggeveen, A., Tsiros, M., & Schlesinger, L. A. (2009). Customer experience creation: Determinants, dynamics and management strategies. *Journal of Retailing*, 85(1), 31–41.

Volland, D., & Eurich, M. (2014). ICT-enabled value creation in community pharmacies: An applied design science research approach. Thirty Fifth International Conference on Information Systems, Auckland 2014.

Wass, S., & Vimarlund, V. (2016). Healthcare in the age of open innovation–A literature review. *Health Information Management Journal*, 45(3), 121–133.

Wayne, G. R. (2012). Open innovation and stakeholder engagement. *Journal of Technology Management & Innovation*, 7(3), 1–11.

Webster, J., & Watson, R. T. (2002). Analyzing the past to prepare for the future: Writing a literature review. *MIS Quarterly*, xiii–xxiii.

Zismer, D. K. (2013). An argument for the integration of healthcare management with public health practice. *Journal of Healthcare Management/American College of Healthcare Executives*, 58(4), 253.

https://doi.org/10.1142/9789811234491_0005

Chapter 5

Innovation in Higher Education: From Contributor to Driver of Internet-Based Service Innovation

Mohammad Ejaz[*,‡] *and Rómulo Pinheiro*[†,§]

[*]*Department of Business, Strategy and Political Sciences Campus Drammen, University College of Southeast Norway, Lærerskoleveien 40, 3679 Notodden, Norway*

[†]*Department of Political Science and Management, University of Agder, Campus Kristiansand, Universitetsveien 25, 4630 Kristiansand, Norway*

[‡]*Muhammad.Ejaz@usn.no*

[§]*romulo.m.pinheiro@uia.no*

Abstract. As largely publicly run and funded institutions, universities are being pressured by their host governments to do more with fewer resources, and thus are being urged to innovate. Given their public character, universities are expected to adopt processes and practices that are transparent and socially accountable, which makes the integration of (open) knowledge from external sources a strategic imperative. That being said, little is yet known about how higher education institutions globally are resorting to open innovation to address increasingly complex technical and institutional environments. This paper contributes the growing literature on open innovation within services and addresses the existing knowledge gap by analysing the development of a virtual mobility master's program at a leading research-intensive university based in the Nordic countries.

Keywords. Open innovation; process innovation; service industries; higher education; virtual learning.

1. Introduction

The majority of developed countries could be classified as having service-dominant economics, since the service sector has overtaken the manufacturing sector in terms of the number of employees and the amount of value creation. Despite services being a major source of employment in developed countries, there are few studies on innovation within the so-called services industry. Services received scholarly attention during the 1980s, but major research projects on innovation in services were only initiated in the 1990s (Miles, 2005). When it comes to studying innovation, most methods of evaluating innovation activities have been followed the technological characteristics of manufacturing industries. However, R&D and patent-based methods are not entirely appropriate and relevant in the case of services (Evangelista & Sirilli, 1995; Djellal & Gallouj, 1999; Love & Mansury, 2007; Mansury & Love, 2008) since service companies are involved in creating an intangible commodity (Miles, 2005) in contrast to the tangible nature of products. For this reason, service industries have often been labeled "innovative laggards" (Windrum & Tomlinsoon, 1999) and seen as users of innovative technology developed by manufacturing industries. This perception emerged and has been strengthened and stabilized due to the fundamental problem of understanding and measuring innovation in services. The heterogeneous nature of services demands research-based knowledge to understand the difference between services and products on the one hand and between different services on the other hand. As the level of research on services has been too scant to be able to compare services with the manufacturing industries, the parameters of measuring service innovation have remained underdeveloped and, in a number of cases, services have been considered part of tangible products. Even though research has been concerned with innovation in high-tech industries, a shift of focus toward services has been gaining momentum for only the last two decades. Studies concentrating on the concepts of "new service development" and "innovations in services" have slowly been appearing. Despite the role that services play in the economies of some developed countries, the pace of momentum in this area is still quite slow. The emerging research studying innovation in services can be placed into three streams, namely assimilation (considers services as part of manufacturing sector), demarcation (attempts to differentiate services on the basis of distinctive characteristics) and synthesis (develops a common framework of analysis for products and services) (Gallouj, 1994; Coombs & Miles, 2004; Gallouj & Windrum, 2009), with the bulk of studies placed in the assimilation stream.

When it comes to study parameters, R&D and patents have been utilized as mechanisms for measuring the level and nature of innovation. Traditional ways of defining R&D do not fit innovation in services. Miles (2007, p. 254) provides an example of the UK's Department of Trade and Industry Guidelines on the definition of R&D for tax return purposes, which state: "[S]cience is the systematic study of the nature and behavior of the physical and material universe. Work in the arts, humanities and social sciences, including economics, is not science for the purpose of these guidelines...". These difficulties have recently been acknowledged, and data from Community Innovation Surveys (CIS) show that services companies have been engaging in R&D activities. They are not adopting traditional R&D mechanisms used by manufacturing industries but are instead involving different departments and project teams in innovation processes (Hipp & Grupp, 2005). Furthermore, R&D is not a guaranteed path to innovation because creation of innovation is not merely a scientific research process (Dosi, 1988). In contrast with manufacturing companies, service companies do not normally rely on patents as the *de facto* mode for protecting intellectual property due to the intangible nature of services (Blind *et al.*, 2003). Service companies apply a number of mechanisms to protect their innovations from imitators, such as patents, trademarks and brands. Even though the research on innovation in services is underdeveloped, some aspects of services have been identified as distinct from products. These aspects show that service firms are dissimilar from manufacturing firms. Tether (2005) analyzed data from the European Innobarometer (an annual survey on activities and attitudes related to innovation) and found differences in patterns of innovation between service and manufacturing firms. Service firms developed innovations through close collaboration with customers and suppliers, while manufacturing firms considered R&D and the flexibility of production methods to be key aspects of innovation. Workforce skills and interactions (Mansury & Love, 2008) are also central to innovation in services. Therefore, more research is needed in the direction of studies that shed light on *demarcation* aspects in order to enrich our understanding of the distinctive nature of innovation in services. This is the remit of this chapter.

While open innovation has diverted attention from the strategic advantages of in-house R&D, investment in building internal capabilities and protection of valuable knowledge, the growing literature on open innovation has nevertheless maintained its primary focus on the manufacturing industries. Insights into open innovation practices in services are scarce (Mina *et al.*, 2014) despite the major contribution of service industries to developed countries' economies. Higher education institutions (HEIs), particularly universities, have been emphasized as major sources of external inputs to innovation (Fabrizio, 2009; Cassiman *et al.*, 2010). Their role is not limited to the provision of scientific knowledge; they also act as major arenas for the transmission of high-level skills and competencies that are critical for economic growth, competitive advantages and the sustainability of firms (Nelson & Romer, 1996; Lester & Sotarauta, 2007; Geiger & Sa, 2008). Collaborations between HEIs and industry are opening up possibilities for the direct involvement of academics in the innovation process. The complex and uncertain nature of the innovation process demands different types of knowledge (Asheim & Coenen, 2005) that are increasingly multi- and trans-disciplinary in nature (Alves *et al.*, 2007). Firms are approaching HEIs in order to confront new challenges and capitalize on new market opportunities (Perkmann *et al.*, 2013). Financial services in particular are benefiting from HEIs, as academics have formally remained part of teams tasked with the development of new services. For example, investment banks have been utilizing external ideas generated by university professors and newly trained PhDs in developing new investment instruments (Chesbrough, 2003). HEIs have long provided valuable support in the development of new products and services to firms that operate in mature industries and do not have the critical resources to invest in internal R&D (Sallbone *et al.*, 1993). University–industry relationships have performed a central role in shaping the industrial innovation process (Perkmann & Walsh, 2007).

Like firms, HEIs are faced with a multiplicity of external demands and increasingly operate in a highly competitive global marketplace (Marginson, 2004). As largely publically run and funded institutions, universities are being pressured by their host governments to do more with fewer resources (Pinheiro *et al.*, 2015), and thus are being urged to innovate and come up with new solutions to both new and existing problems (Pinheiro *et al.*, 2018). What is more, given their public character, universities are expected to adopt processes and practices that are easily accountable, which makes the integration of (open) knowledge from external sources a strategic imperative in contemporary HEI management (Pinheiro, 2016). That being said, little is yet known about how HEIs globally are resorting to open innovation as a means of addressing increasingly complex technical and institutional environments. Thus, this chapter contributes to, first, the growing literature on open innovation within services and, second, the knowledge gap within the field of higher education by analysing the development of a virtual mobility master's-level program at a leading research-intensive university based in the Nordic countries. In so doing, we address the following research question:

> *How can the innovation process in higher education be characterized, and how does it evolve over time?*

Sections 2 and 3 provide a brief review of the existing literature in the realm of process and open innovation. They are followed by the methodological considerations underpinning the study in Section 4. Section 5 provides the backdrop for the empirical case and sketches out the key findings from the study. Section 6 discusses the findings in the light of the extant literature and reflects on the key contributions of the study for a better understanding of process innovation within public educational services.

2. The Process Dimension of Innovation: An Evolutionary Perspective

Schumpeter (1934) was the first to shed light on the process of invention in the context of economic growth. Process views on innovation were given an explicit direction by the combination of Schumpeter's work with Usher's (1954) cumulative synthesis theory and seminal insights from Ruttan (1969). According to their line of reasoning, innovation is a process of emergence of new things and solutions in diverse settings. Although the process nature of innovation was thereafter accepted and emphasized, the general assumption continued to revolve around the smooth and linear nature of the process, which started with basic research and ended in the form of products and was generally labeled the linear model of innovation (Santamaría *et al.*, 2009). Kline and Rosenberg (1986) initiated the move from linearity to uncertainty reflecting an increased complexity of the system both technically and socially. Their seminal work not only provided a systemic direction to innovation-process research but, equally important, inspired many other researchers engaged in studying innovation. Kline and Rosenberg's point of departure was to assume that, as a process, innovation results from the combination of diverse ideas and solutions that are themselves driven by technologic and economic rationales. That is, the process proceeds because of internal and external feedback. When existing and available knowledge does not address the problems that emerge during an innovation process, research is conducted to create new knowledge to address such problems.

In pursuing the process line of reasoning, Nonaka and Kenney (1991) studied the product development process at two companies (Apple and Canon) and found innovation to be an information creation process in which human actors are labeled information creators and information processors. The process produces two types of information: *syntactic* and *semantic*, with the former capable of being converted into digital form and having no inherent meaning and the latter being qualitative, holistic, meaningful and transformational. The creation of semantic information was seen to be the result of new insights rather than deductive modeling. Leaders carefully managed information through the creation of an appropriate organizational structure and via transformational mechanisms. They selected relevant personnel and helped them to overcome barriers and challenges that emerged during the innovation process.

Van de Ven *et al.*'s (1999) work illuminated the complexity of events and, subsequently, the non-linearity of the innovation process in start-ups, internal corporate ventures and joint inter-organizational ventures. The innovation process was then defined "as a non-linear cycle of divergent and convergent phases of activities that may be repeated over time and at different organizational levels if resources are obtained to renew the cycle" (Van de Ven *et al.*, 1999, p. 184). *Divergence* is related to the expansion of dimensions and the chaotic nature of the process while *convergence* refers to a more directed and narrow process in which the locus of attention is the implementation of ideas and strategies.

As a research field, innovation studies have undergone change and transformation in recent years that could be attributed to the multidisciplinary nature of the field. Researchers from a number of areas, such as economics, sociology, psychology and management sciences, have been enriching the field with diverse knowledge. One of the more debated topics in the field of innovation emerged in the beginning of the 2000s as a result of scholarly work on *open innovation*, where knowledge is purposively allowed to flow in and out of an organization in a bid to accelerate the innovation process. Chesbrough and Crowther (2006) described open innovation as following a historical tradition of approaching innovation as a process that entails "invention implemented and taken to market" (Chesbrough, 2003, p. ix). Hence, the creation of an idea cannot be labeled innovation if it does not pass through a process intended to make desired changes and it is not adopted by users.

3. The "Outside–In" Dimension of Open Innovation

Outside–in exploitation of knowledge has received enormous attention from open innovation researchers in the last decade. Individuals, customers, suppliers, competitors and knowledge organizations like universities are some of the sources providing complementary inputs to innovation processes in a number of industries. The so-called asset-intensive industries (Chesbrough & Crowther, 2006), small- and medium-sized enterprises (van de Vrande *et al.*, 2009; Brunswicker & Vanhaverbeke, 2015) and service industries (Mina *et al.*, 2014) are the major recipients of externally generated knowledge. These industries are capitalizing on open innovation to accelerate their innovation capabilities and enhance performance. However, the exploitation of knowledge does not entail a free ride on the discoveries of others who created research, inventions and innovations through the utilization of valuable resources. Appropriability and the protection of knowledge are strongly emphasized, advocated and promoted in the era of globalization, digitalization and the so-called sharing economy, when phenomena such as open source are normally blended with open innovation (Chesbrough, 2011). The role played by HEIs has long been elaborated within the large stream of research involving university–industry collaborations, academic spin-offs, science parks, and so on (Perkmann *et al.*, 2013; van Geenhuizen & Soetanto, 2009; Saublens *et al.*, 2016). Both manufacturing and service industries have benefited from critical knowledge generated through basic research at HEIs (Owen-Smith *et al.*, 2002; Hauge *et al.*, 2016). As such, the knowledge emanating from academia has contributed to the innovation process for products and services. Despite the widespread consensus that HEIs are critical for business and social innovations, the development of regional innovation systems and local attraction of both high-level skills and venture capital, these vital institutions are scarcely recognized as creators of innovation. Moreover, given the sweeping changes occurring within the larger field of higher education (as sketched out above), an argument could be made that HEI involvement in the innovation process is not restricted to an inside–out contribution, and this fact is broadly recognized by the extant literature. For example, the rise of massive open online learning services or MOOCs has brought to the fore a set of new opportunities but also challenges for HEIs (Tømte *et al.*, 2020).

For the last couple of decades, radical developments in information and communication technologies (ICT) have contributed substantially to efforts associated with searching, identifying and integrating knowledge. Nevertheless, organizations confront both internal and external challenges, risks, complexities and uncertainties in innovation activities. For example, on the path towards collaboration with other partners, organizations often face internal resistance to external knowledge, which is labeled the "not-invented-here" syndrome (Katz & Allen, 1982). Building internal absorptive capabilities, being equipped with knowledge of partners' operations and articulating negotiation strategies are fundamental for the effective and smooth utilization of knowledge at the organizational level.

4. Methodology

This research work was designed to provide insights into dynamic processes of service innovation in higher education, a key sector for national economies in transition to a knowledge-based society and economy (Temple, 2011). Methodologically, we have chosen a case study research design as a way of gaining critical empirical insights into the process of service development within an HEI context. Case studies allow us to explore a given phenomenon in its real-life context and where the boundaries between phenomena (i.e., a new degree program) and context (the higher-education field) are not clearly demarked (Yin, 2009). Even though case study strategies have been followed in a number of previous studies focusing on open innovation, the majority of these used organizations (meso level) as the unit of analysis (Du *et al.*, 2014). In this study, the focus is shifted from the meso to the micro level of the project (a master's program) as the primary unit of analysis. Due to its novel approach, the virtual mobility program (see below) hosted at the University of Oslo was selected as the case for this study. The researchers involved did not have any influence or prior knowledge of the objectives, decisions and direction of the process being studied. The path to research was triggered by an unplanned encounter with the responsible professor at the case university. This first face-to-face informal discussion in the spring of 2017 did not involve any specific target research area, but focused mostly on the collaboration with other partners and the nature and scope of educational and research endeavours. Nevertheless, this initial discussion provided the basis for unearthing the innovation process associated with the master's program that is the unit of analysis. This informal mode of becoming equipped with valuable information was very helpful in setting the research focus, understanding the rationales behind the establishment of a new mode of internationalization and identifying potential sources of data collection (triangulation).

The formal data collection process started with observation of the service provision in practice. In spring 2017, one of the researchers participated as an observer and attended a teaching and discussion session along with students. This mechanism of data collection provided critical insights into the functioning of the technology, teaching and learning activities, as well as into the interaction amongst the different participants. After class participation, a face-to-face interview was conducted with a university staff member who played a leading role in recommending and organizing technological solutions at the faculty level. This interview lasted for more than one hour and was recorded. Hence, two interviews and two hours of participation in classroom settings as an observer provided the basis for the data collection. The interviews were transcribed verbatim and were analyzed and coded using the NVivo 10 software package.

5. Empirical Section

5.1. *Contextual backdrop to the case*

Founded in 1811 as the Royal Frederick University, the University of Oslo (UiO) has a long and established history of providing educational services and undertaking high-level research activities. As the oldest and most prestigious centre for teaching, learning and research in Norway, it has been continuously performing a key role catering to the growing demand for skilled labour in both the industrial and the knowledge-based economies. Following the implementation of a quality reform in 2003 (Stensaker *et al.*, 2005), the university actively embarked on a path of internationalization and offered new directions for teaching and research through its active participation in various mobility programs, such as Erasmus Mundus, and facilitation of staff/research mobility. By riding the tide of internationalization that swept through the whole of Europe at the turn of the century (Gornitzka & Langfeldt, 2008), UiO emerged as a key participant in the international dimension of knowledge. The adoption of a new approach to teaching, learning and mobility paved the way for a gradual increase in the portfolio of programs being offered, the recruitment of international students on all levels and the number of researchers (many of them foreign born) involved.

Among the many "new" master's programs offered by UiO, the higher education program (HEP) hosted by the Faculty of Education (FE) is one of the older ones taught in the English language, going back to the early 2000s. The program has remained a key asset within a portfolio of programs that strive for the promotion of internationalization within both the faculty and the university as a whole. Moreover, it has continuously transformed its teaching, learning and mobility practices. Lately, the program embarked on a journey of service innovation termed "virtual mobility" in an effort to capitalize on new opportunities offered by the emergence and diffusion of ICT. Through virtual mobility, in-house students enrolled at HEP and graduate students based at the College of Education at the University of Iowa in the United States attend common courses supported by a technology called "Zoom". A description of the innovation process associated with this "virtual mobility" service from idea creation to implementation is succinctly described below.

5.1.1. *Idea creation*

The master's program in higher education studies has a strong focus on collaboration with external partners in the identification, creation and provision of educational and training services. The defining characteristic of a collaborative approach capitalizes on distributed resources and competencies and the international nature of knowledge in which multiple educational partners from a number of countries serve a broad range of objectives. These partners are not only valuable sources of external knowledge but also perform the critical function of hosting outgoing students and researchers at their institutions. When it comes to searching, identifying and integrating knowledge, geographic boundaries create no restrictions. The program utilizes the knowledge and academic capabilities (i.e., guest professors) of people based in different countries and working for a number of diverse HEIs. The organizers of the program realized the need to embark on a journey of innovation through riding the tide of digitalization. They came up with the idea of having "visual mobility" through real-time online communication with a group of students and researchers based at the University of Iowa. It was an attempt to provide every student with the opportunity for mobility, interaction and participation, which is difficult in the traditional mode of internationalization. Although many students benefited from the existing student mobility opportunities offered at UiO, most were unable to participate in internationalization due to the scarcity of resources supporting mobility and other associated challenges. Hence, it was hypothesized that a new HEI service in the form of visual mobility could be more successful as international trends moved in the direction of internet-based educational services based on either open (MOOCS) or closed models.

To transform the original idea into a functional service, a project proposal was prepared to tap into a new funding scheme under the auspices of Norway's Centre for International Cooperation in Education (SIU). This new mechanism of interaction and mobility was expected to offer new directions for internationalization practices in Norway. However, the actual building of the new technological platform was not part of the project since it has been anticipated that available technological solutions at UiO's FE would address future educational services. In fact, the FE was already equipped with state-of-the-art ICT-based infrastructure and know-how aimed at facilitating interaction among students and researchers based at different universities and located on different continents.

5.1.2. *Development of the technological platform*

After securing financial resources from SIU, the program-level leadership decided to assess the availability of technological capabilities at FE. Since over the last two decades almost all sectors of national economies have been transforming to digitalization, it was anticipated that FE would be equipped with the most advanced technological equipment. The technical manager was given the responsibility to evaluate the viability of existing ICT solutions since earlier estimates had been based on mere assumptions. The FE had a so-called electronics teaching room that was well organized and decorated. The faculty's technical manager was fascinated by the presence of many devices when he stepped into the room. The room had many diverse types of devices that could solve technological challenges. However, it was soon realized that the majority of devices were not in working order. Despite the expenditure of so many financial and human resources, no one had used the room for teaching and learning purposes in the previous six months.

Given the failure of search for a suitable classroom and the non-availability of any other options at FE, the technical manager was faced with the need to find a new solution at another department or institution, regardless of the fact that the faculty had been a leading player in shaping the teaching and learning environment. However, there was another electronics room at the library that was greatly appreciated by a number of professionals. The technical manager decided to assess whether this room could provide the necessary technological solution and an arena for teaching and learning activities. However, once again, the room was filled with old devices that were unable to fulfil the requirements of real-time interaction.

The non-availability of any viable technological solution at UiO led to the evaluation of alternative resources owned by the collaboration partner overseas. The technical manager, along with another researcher, decided to travel to Iowa in January 2017. Founded in 1847, the University of Iowa (UoI) had developed a digital learning platform that was facilitating communication among a broad range of stakeholders. It was designed as a simple technical platform that could handle multiple tasks simultaneously. This platform allowed students and teachers a great deal of freedom in communicating with each other. Students could communicate with teachers and with fellow students in a virtual class. Since UoI had gained significant expertize in running this platform, it was realized that the platform could perform a vital role in virtual mobility. A single electronics device with multiple cameras could provide state-of-the-art and excellent quality audio and video coverage and communication among students based at two geographic locations through a software solution called "Zoom". Therefore, it was decided to purchase the devices, as all available technological devices and options at UiO were unable to fulfil the objectives of the new service. Nevertheless, finding a suitable technological solution was not the only challenge that the new program confronted. Protection of personal data was necessary if students were to interact purposefully with other students and researchers without compromising their personal identities and intellectual capital. The new mode of creating a HEI service was supposed to bring together students from two different programs having separate technological platforms with diverse user names and passwords. To confront this challenge, the technical managers of the partner universities decided to create a website that not only became a new closed common software platform for students and researchers, but also provides necessary open access and information for the general public. After developing and testing a set of ICT-based fundamental solutions, the program-level leadership and the technical managers of all involved partners made the decision to offer the new service to their regular master's-level students.

5.1.3. *The offering of the service*

The new educational service was launched on 13 February 2017. Students in both master's programs participate in common classroom settings with shared slides and shared presentations. The interaction between the two different groups of students is supported by two audio/video devices at two different locations. Each device consists of five cameras covering a 360° and transmitting audio and video in real time. The salient features of these cameras are autonomous, automatic and sound-focused coverage without any additional need for diversion. However, the five different cameras do not send audio/video transmissions simultaneously, as the cameras follow the sounds of the speakers. Upon the completion of one speaker's comment, a second camera starts covering another speaker. This mode of coverage helped solve the challenge of broadcasting multiple transmissions when more than one speaker is actively participating in discussions. Furthermore, participants have the opportunity to create a dialogue-based environment in which all participants can take part in discussions. Students have formed groups for discussions and assignments without taking into consideration the geographic distance (a 7-h time zone difference) between the two locations.

6. Discussion and Conclusions

The concept of open innovation has been the focus of widespread attention by both academics and practitioners since the appearance of Chesbrough's ground-breaking work. However, only a few studies have shed light on open innovation in the specific context of HEI services. This current study attempts to enrich understanding of the journey of innovation in one of the more important sectors for national economies. Our analysis shows that HEIs integrate and utilize external knowledge in innovation processes through outside–in activities. Their search for knowledge is not limited to regional actors; they collaborate with partners based on other continents. The distance between the partners' locations is not a hindrance since ICT has transformed ways of searching and integrating knowledge and competencies. In the integration of knowledge, the technological capabilities of the knowledge provider set the direction of technology. The creation of innovation based on tested technological solutions reduces the time required to introduce services to users, thus enhancing efficiency. Familiarity with technology helps overcome many uncertainties commonly associated with innovations (Chesbrough, 2004). HEIs such as the chosen case are not utilizing technology in a standardized form (of technology-based services) since service creators require a great deal of transformation in the design and delivery of services. The design and provision of services are based on an open mode, as service providers have the flexibility of choosing both the technology and the content. Technological solutions can easily be replaced with new solutions, but transformation involves extra costs in the form of staff training and the integration of new and old technologies.

In relation to users' participation, the findings from our single case study indicate that HEIs are not actively engaging customers or users (i.e., students) in the innovation process. This finding is contradictory to previous research on open innovation that highlights customers or users as the most important source of external knowledge (Grimpe & Sofka, 2009; Gassmann, 2006)). The lack of customers' or users' involvement in the key phases of ideation, design, implementation and testing might be due to the new mode of application of ICT in teaching and learning. Even though ICT has been performing a complementary role in teaching and learning, it has emerged as a central and necessary mode of communication in the context of virtual mobility. In this new role, the continuous participation of customers or users could reduce the challenge of delays in the adoption of new services. Users are equipped with knowledge that could help in the identification of their own needs. Some of these needs are difficult to analyze beforehand and require the active participation of users in the innovation process (Lundvall, 1985). The gradual testing of services in practice is commonly associated with new services and product development since those who create innovations cannot anticipate the ways in which services will be utilized in the real word (Andersen, 1991). Furthermore, producers are also unable to predict the value these new services will create or the potential interests of their users (Bogers *et al.*, 2010). Compared with traditional classroom-based interaction among academics and students, the new modes of mobility utilize technological and human resources in completely different settings. In the traditional mode of communication, teachers and students directly exchange their views, experiences and interests, both inside the classroom setting and within the broader context of the physical campus. However, the rise of virtual mobility has opened up new arenas for the exchange of knowledge between academics and students on the one hand, and amongst students located at different institutions and locations on the other hand.

Our case findings highlight the fact that, as knowledge-laden organizations operating in complex institutional and technical environments (Pinheiro & Young, 2017), the transfer of knowledge amongst modern universities occurs, inter alia, through formal, external networking (van de vrande *et al.*, 2009) in which both partners have a strong commitment to service innovation. On the path towards innovation, HEIs have been found to rarely utilize the breath and depth of possible search strategies (Laursen & Salter, 2006), that is, to search for partners that are well equipped with human, financial and knowledge-related resources. As is the case in other educational settings, such as joint research projects, personal and individual relationships provide both the basis for selecting collaborative partners and for building formal, strong organizational networks that lead to the introduction of innovations based on similarities in existing services. It is widely known that academics in Norway and beyond are generally involved in strong individual networks that span organizational and regional boundaries (Gornitzka & Langfeldt, 2008). These networks are global in nature since researcher and student mobility is strongly promoted by HEI policies all over the world (Jacob & Meek, 2013). Individual (informal) networks, in turn, provide the foundations for establishing formal organizational networks such as joint programs and activities. As a result of the establishment of strong organizational networks, HEIs are creating educational services by combining scattered human, financial and knowledge-based resources. Without engagement through strong networks, the new combination of resources to create innovations could pose many challenges, since the users of HEI services utilize such services continuously. As the Internet transforms the ways in which service providers and user groups interact, the distribution of valuable, tacit knowledge across national and organizational borders through novel services with intangible characteristics is steadily becoming a reality. Stated differently, the Internet is transforming academic networks and opening up possibilities for launching new, innovative services. This, in turn, is allowing HEIs to respond more efficiently to the challenges posed by increasingly competitive national, regional and global environments in their quest to become both more entrepreneurial and world class (Geschwind & Pinheiro, 2017).

Future studies, ideally using both qualitative and quantitative methodologies, could provide further empirical and theoretical accounts on the ways in which HEIs, in the Nordics and beyond, and other types of knowledge-based public organizations for that matter, engage in innovative efforts within the scope of an open/service innovation framework. What is more, longitudinal accounts would allow us to take stock and make sense of such developments over time, hence shedding critical light on the interplay between the multiple stages of the innovation process, from idea generation to implementation to rejection or de-institutionalization.

References

Alves, J., Marques, M. J., Saur, I., & Marques, P. (2007). Creativity and innovation through multidisciplinary and multisectoral cooperation. *Creativity and Innovation Management*, 16(1), 27–34.

Andersen, E. S. (1991). Techno-economic paradigms as typical interfaces between producers and users. *Journal of Evolutionary Economics*, 1(2), 119–144.

Asheim, B. T., & Coenen, L. (2005). Knowledge bases and regional innovation systems: Comparing Nordic clusters. *Research Policy*, 34(8), 1173–1190.

Blind, K., Edler, J., Schmoch, U., Andersen, B., Howells, J., Miles, I., Roberts, J., Green, L., Evangelista, R., & Hipp, C. (2003). *Patents in the Service Industries* (final report prepared for the European Commission). Fraunhofer ISI, Karlsruhe.

Bogers, M., Afuah, A., & Bastian, B. (2010). Users as innovators: A review, critique, and future research directions. *Journal of Management*, 36(4), 857–875.

Brunswicker, S., & Vanhaverbeke, W. (2015). Open innovation in small and medium-sized enterprises (SMEs), External knowledge sourcing strategies and internal organizational facilitators. *Journal of Small Business Management*, 53(4), 1241-–1263.

Cassiman, B., Chiara Di Guardo, M., & Valentini, G. (2010). Organizing links with science: Cooperate or contract?: A project-level analysis. *Research Policy*, 39(7), 882–892.

Chesbrough, H. (2003). *Open innovation: The new imperative for creating and profiting from technology*. Harvard Business Press.

Chesbrough, H. (2004). Managing open innovation. *Research-Technology Management*, 47(1), 23–26.

Chesbrough, H. W. (2011). Bringing open innovation to services. *MIT Sloan Management Review*, 52(2), 85.

Chesbrough, H., & Crowther, A. K. (2006). Beyond high tech: Early adopters of open innovation in other industries. *R&D Management*, 36(3), 229–236.

Coombs, R., & Miles, I. (2000). Innovation, measurement and services: The new problematique. In: J. S. Metcalfe and I. Miles (eds.), *Innovation Systems in the Service Economy*. Kluwer, Boston, pp. 85–103.

Djellal, F., & Gallouj, F. (1999). Services and the search for relevant innovation indicators: A review of national and international surveys. *Science and Public Policy*, 26(4), 218–232.

Dosi, G. (1988). The nature of the innovative process. In: G. Dosi, C. Freeman, and R. Nelson (eds.), *Technical Change and Economic Theory*. Pinter Publishers, London, pp. 221–238.

Du, J., Leten, B., & Vanhaverbeke, W. (2014). Managing open innovation projects with science-based and market-based partners. *Research Policy*, 43(5), 828–840.

Evangelista, R., & Sirilli, G. (1995). Measuring innovation in services. *Research Evaluation*, 5(3), 207–215.

Fabrizio, K. R. (2009). Absorptive capacity and the search for innovation. *Research Policy*, 38(2), 255–267.

Gallouj, F. (1994). *Economie de l'Innovation dans les Services*, Editions L'Harmattan, Logiques Économiques, Paris. (English language version: Gallouj, F. (2002). *Innovation in the Service Economy: The New Wealth of Nations*). Edward Elgar, Cheltenham.

Gallouj, F., & Windrum, P. (2009). Services and services innovation. *Journal of Evolutionary Economics*, 19(2), 141–148.

Gassmann, O. (2006). Opening up the innovation process: Towards an agenda. *R&D Management*, 36(3), 223–228.

Geiger, R. L., & Sá, C. M. (2008). *Tapping the Riches of Science: Universities and the Promise of Economic Growth*. Harvard University Press, Boston.

Geschwind, L., & Pinheiro, R. (2017). Raising the summit or flattening the agora? the elitist turn in science policy in Northern Europe. *Journal of Baltic Studies*, 48(4), 513–528.

Gornitzka, Å., & Langfeldt, L. (2008). *Borderless Knowledge: Understanding the "New" Internationalisation of Research and Higher Education in Norway*. Springer, Dordrecht.

Grimpe, C., & Sofka, W. (2009). Search patterns and absorptive capacity: Low- and high-technology sectors in European countries. *Research Policy*, 38(3), 495–506.

Hauge, E. S., Pinheiro, R. M., & Zyzak, B. (2018). Knowledge bases and regional development: Collaborations between higher education and cultural creative industries. *International Journal of Cultural Policy*, 24(4), 485–503.

Hipp, C., & Grupp, H. (2005). Innovation in the service sector: The demand for service-specific innovation measurement concepts and typologies. *Research policy*, 34(4), 517–535.

Jacob, M., & Meek, V. L. (2013). Scientific mobility and international research networks: Trends and policy tools for promoting research excellence and capacity building. *Studies in Higher Education*, 38(3), 331–344.

Katz, R., & Allen, T. J. (1982). Investigating the not invented here (NIH) syndrome: A look at the performance, tenure, and communication patterns of 50 R&D project groups. *R&D Management*, 12(1), 7–20.

Kline, J. K., & Rosenberg, N. (1986). An overview of innovation. In: R. Landau and N. Rosenburg (eds.), *The Positive Sum Strategy: Harnessing Technology for Economic Growth*. National Academy Press, Washington DC, pp. 275--386.

Laursen, K., & Salter, A. (2006). Open for innovation: The role of openness in explaining innovation performance among UK manufacturing firms. *Strategic Management Journal*, 27(2), 131–150.

Lester, R., & Sotarauta, M. (Eds.). (2007). *Innovation, Universities and the Competitiveness of Regions*. Tekes, Helsinki.

Love, J. H., & Mansury, M. A. (2007). External linkages, R&D and innovation performance in US business services. *Industry and Innovation*, 14(5), 477–496.

Lundvall, B. Å. (1985). Product innovation and user-producer interaction. *The Learning Economy and the Economics of Hope*. Antham Press, London and New York.

Mansury, M. A., & Love, J. H. (2008). Innovation, productivity and growth in US business services: A firm-level analysis. *Technovation*, 28(1), 52–62.

Marginson, S. (2004). Competition and markets in higher education: A "glonacal" analysis. *Policy Futures in Education*, 2(2), 175–244.

Miles, I. (2005). Innovation in services. In: J. Fagerberg, D. Mowery, and R. Nelson (eds.), *The Oxford Handbook of Innovation*. Oxford University Press, Oxford and New York, pp. 433–458.

Miles, I. (2007). Research and development (R&D) beyond manufacturing: The strange case of services R&D. *R&D Management*, 37(3), 249–268.

Mina, A., Bascavusoglu-Moreau, E., & Hughes, A. (2014). Open service innovation and the firm's search for external knowledge. *Research Policy*, 43(5), 853–866.

Nelson, R. R., & Romer, P. M. (1996). Science, economic growth, and public policy. *Challenge*, 39(1), 9–21.

Nonaka, I., & Kenney, M. (1991). Towards a new theory of innovation management: A case study comparing Canon, Inc. & Apple Computer, Inc. *Journal of Engineering and Technology Management*, 8(1), 67–83.

Owen-Smith, J., Riccaboni, M., Pammolli, F., & Powell, W. W. (2002). A comparison of U.S. and European university–industry relations in the life sciences. *Management Science*, 48(1), 24–43. doi:10.1287/mnsc.48.1.24.14275.

Perkmann, M., & Walsh, K. (2007). University–industry relationships and open innovation: Towards a research agenda. *International Journal of Management Reviews*, 9(4), 259–280.

Perkmann, M., Tartari, V., McKelvey, M., Autio, E., Broström, A., D'Este, P., . . . Sobrero, M. (2013). Academic engagement and commercialisation: A review of the literature on university–industry relations. *Research Policy*, 42(2), 423–442.

Pinheiro, R. (2016). Humboldt meets Schumpeter? interpreting the "entrepreneurial turn" in European Higher Education. In: S. Slaughter and B. J. Taylor (eds.), *Competitive advantage: Stratification, privatization and vocationalization of Higher Education in the US, EU, and Canada*. Springer, Dordrecht, pp. 291–310.

Pinheiro, R., & Young, M. (2017). The university as an adaptive resilient organization: A complex systems perspective. In: J. Huisman and M. Tight (eds.), *Theory and Method in Higher Education Research*. Emerald, Bingley, pp. 119–136.

Pinheiro, R., Wengenge-Ouma, G., Balbachevsky, E., & Cai, Y. (2015). "The role of higher education in society and the changing institutionalized features in higher education." In: J. Huisman, H. de Boer, D. Dill, and M. Souto-Otero (eds.), *The Palgrave International Handbook of Higher Education Policy and Governance*. Palgrave Macmillan, London and New York, pp. 225–242.

Pinheiro, R., Mitchell, Y., & Šima, K. (2018). *Higher Education and Regional Development: Tales from Northern and Central Europe*. Palgrave, Cham.

Ruttan, V. W. (1959). Usher and Schumpeter on invention, innovation, and technological change. *The Quarterly Journal of Economics*, 73(4), 596–606.

Santamaría, L., Nieto, M. J., & Barge-Gil, A. (2009). Beyond formal R&D: Taking advantage of other sources of innovation in low-and medium-technology industries. *Research Policy*, 38(3), 507–517.

Saublens, C., Bonas, G., Husso, K., Komárek, P., Oughton, C., & Santos Pereira, T. (2016). *Regional Research Intensive Clusters and Science Parks*. European Commission, DG Research, Brussels.

Schumpeter, J. (1934). *The Theory of Economic Development: An Inquiry into Profits, Capital, Credit, Interest, and the Business Cycle*. Transaction Publishers, New Brunswick, New Jersey.

Stensaker, B., Aamodt, P., Arnesen, C., Mariusen, Å., Fraas, M., Gulbrandsen, M., ... Olsen, T. (2005). *OECD Thematic Review of Tertiary Education: Country Background Report for Norway*. Oslo: The Ministry of Education and Research/NIFU-STEP, Oslo.

Temple, P. (2011). *Universities in the Knowledge Economy: Higher Education Organisation and Global Change*. Taylor & Francis, London.

Tether, B. S. (2005). Do services innovate (differently)? Insights from the European Innobarometer Survey. *Industry & Innovation*, 12(2), 153–184.

Tømte, C., Laterza, V., & Pinheiro, R. (2020). Is there a Scandinavian model for MOOCs? Understanding the MOOC phenomenon in Denmark, Norway, and Sweden. *Nordic Journal of Digital Literacy*, 15(4), 234–245.

Usher, A. P. (1954). *A History of Mechanical Inventions*. Harvard University Press, Cambridge.

Van de Vrande, V., De Jong, J. P. J., Vanhaverbeke, W., & De Rochemont, M. (2009). Open innovation in SMEs: Trends, motives and management challenges. *Technovation*, 29(6), 423–437.

van Geenhuizen, M., & Soetanto, D. P. (2009). Academic spin-offs at different ages: A case study in search of key obstacles to growth. *Technovation*, 29(10), 671–681. doi: 10.1016/j.technovation.2009.05.009

Windrum, P., & Tomlinson, M. (1999). Knowledge-intensive services and international competitiveness: A four country comparison. *Technology Analysis & Strategic Management*, 11(3), 391–408.

Yin, R. K. (2009). *Case Study Research: Design and Methods*. Sage Publications, London.

https://doi.org/10.1142/9789811234491_0006

Chapter 6

Intra-bound Innovation and Strategizing in Service MNCs

Katja Maria Hydle[*,‡] *and Kristin Wallevik*[†,§]

[*]*Innovation, Society, NORCE Norwegian Research Centre, Essendropsgate 3, 0368 Oslo, Norway*

[†]*School of Business and Law, University of Agder, Box 422, 4604 Kristiansand, Norway*

[‡]*katja.hydle@norceresearch.no*

[§]*kristin.wallevik@uia.no*

Abstract. This chapter explores open innovation and explicit and emergent strategizing in Norwegian and foreign-based multinationals in the oil and gas service sectors. The chapter combines literature on open innovation and strategizing to shed light on mechanisms within firms regarding openness in innovation. Findings from 10 multinational companies demonstrate that company acquisition and location sourcing are explicit strategizing activities based at headquarters and expose inbound innovation in line with existing literature. Everyday practices on the periphery show that activities acquisition and knowledge sourcing are patterns of emergent strategizing that are performed locally and expose *intra-bound innovation*, nuancing existing open innovation categories. Our findings also nuance the understanding of strategizing from headquarters by demonstrating ambidexterity of both center and periphery. Strategizing is both centralized and dispersed and there are inbound and intra-bound innovation in use between the headquarters and the periphery that are initiated and evolve from both places, complementing each other. By combining research on strategy with open innovation, we expose center and periphery ambidexterity and provide open innovation research with an understanding of push and pull forces for intra-bound innovation.

Keywords. Open innovation; explicit and emergent strategizing; oil and gas service sector; company acquisition; location sourcing; activities acquisition; knowledge sourcing; intrabound innovation.

1. Introduction

Open innovation "is a distributed innovation process that relies on purposively managed knowledge flows across organizational boundaries, using pecuniary and non-pecuniary mechanisms in line with the organization's business model to guide and motivate knowledge sharing" (Chesbrough, 2017). Opening up innovation processes through "the use of purposive inflows and outflows of knowledge" is aimed at enhancing internal innovation processes and accessing markets while reducing innovation costs (Chesbrough, 2003, 2012). However, open innovation research (Dahlander & Gann, 2010) calls for research into how firms' strategies are operationalized in order for them to benefit from openness in innovation, including the various mechanisms to obtain this openness and how resources are enacted. This chapter aims to answer this call by looking at service MNCs and their strategizing for innovation.

Open innovation and internal innovation activities are found to be part of firms' strategizing (Hydle, 2014, 2015). Hydle (2015) finds that emergent strategizing in MNCs occurs through innovation activities, which in turn frames future strategizing. Innovation is closely linked to future and long-term value creation. For this reason the enhancement of innovation activities are found to be strategic (Hydle, 2015). This study mainly focuses on emergent strategizing. Most strategy-related research has focused on explicit strategizing activities and less on strategizing that is emergent, related to coping practices or inductive strategizing (Chia & MacKay, 2007; Jarzabkowski *et al.*, 2007; Jarzabkowski & Spee, 2009; Kornberger & Clegg, 2011; Vaara & Whittington, 2012; Tsoukas, 2010; Mintzberg & Waters, 1985; Chia & Holt, 2006). Few studies focus on both explicit and emergent strategies in MNCs. Regnér (2003), however, finds that emergent strategies play an important role at the periphery of organizations, where strategy is developed through externally-oriented, explorative activities. Established strategy from the center, on the other hand, is more industry-focused, emphasizing exploitation. Hence, strategizing at the periphery is understood as inductive, while strategizing at the center is found to be deductive (Regnér, 2003). This chapter builds on these insights, exploring both explicit strategizing and emergent strategizing in parallel (Hydle, 2015; Vaara & Whittington, 2012; Regnér, 2003). The main research question is thus: *How and why do strategizing activities occur in MNCs in relation to open innovation?* We combine literature on open innovation and strategizing to shed light on MNCs within the oil and gas service sector.

This chapter reports a collective case study of 10 MNCs in the oil and gas industry in Norway, focusing on innovation activities, both from HQs and subsidiaries. The services from these MNCs are engineering services related to oil and gas operators and projects, but also services to the oil and gas industry in general. In this industry, there are a multitude of specialized services that can either be standardized or customized, depending on the need. Some of these services are: project coordination, management, third-party services, engineering analysis, detailed engineering, seismic services, operation execution, site and logistics management, material procurement, field development services and consultancy.

Through our study, four strategizing practices of innovation are identified. First, *company acquisitions* are strategically used to gain access to innovation and to distribute the innovation within the firm, globally. Second, *location sourcing* of subsidiaries is used strategically to build up local innovation activities. Third, *activities acquisitions* are everyday practices where innovation activities gradually take form. Fourth, *knowledge sourcing* involves everyday practices in which activities are performed where knowledgeable employees are located. The contribution of this study is to identify how strategizing is both centralized and dispersed in parallel, and that MNCs intra-bound innovation approach encompasses a dynamic relationship between headquarters and the subsidiary, formed and driven from both the center and the periphery.

The chapter proceeds with a theoretical understanding of open innovation and the building and dwelling modes of strategizing (Chia & Holt, 2006; Chia & Rasche, 2010). In the ensuing section, a collective case study of 10 MNCs is introduced, followed by an exposition of the innovation activities between HQ and subsidiaries, and an analysis of how these innovation activities are linked to explicit or emergent strategy. This is discussed in relation to strategizing and the framework of open innovation.

2. Theoretical Framing

2.1. *Open innovation*

In open innovation research, a distinction is made between inbound and outbound innovation (Dahlander & Gann, 2010; Chesbrough & Crowther, 2006; Bianchi *et al.*, 2011). Inbound open innovation is about leveraging the technologies and discoveries of others, implying "opening up" to others to obtain this. Outbound open innovation is about establishing relationships with external others to transfer technologies for commercial exploitation. In a review of open innovation research, Dahlander and Gann (2010) further divide inbound and outbound innovation to include interactions that are pecuniary vs. non-pecuniary. By doing this, they expose four different categories of open innovation (Table 1).

Table 1. Dahlander and Gann (2010) different forms of openness.

	Inbound innovation	Outbound innovation
Pecuniary	Acquiring	Selling
Non-pecuniary	Sourcing	Revealing

Pecuniary inbound innovation is about "*acquiring* input to the innovation process through the market place ... how firms license-in and acquire expertise from outside" (Dahlander and Gann, 2010). Non-pecuniary inbound innovation, refers to how *external sources* of innovation can be used internally. Furthermore, pecuniary outbound innovation is about *how firms sell* or license out inventions, technologies and resources that have been developed in the firm. Lastly, non-pecuniary outbound innovation refers to how internal resources are *revealed externally* (Dahlander and Gann, 2010). The review discusses advantages and disadvantages of each form of openness, acknowledging that there is limited understanding of the sourcing processes of external knowledge into corporations, as well as little understanding of how operational strategies are used, to benefit from open innovation. We thus turn to literature on strategizing to view how open innovation mechanisms are operationalized in firms.

2.2. *Strategizing*

Strategizing is understood as "the myriad of activities that lead to the creation of organizational strategies. This includes strategizing in the sense of more or less deliberate strategy formulation, the organizing work involved in the implementation of strategies, and all the other activities that lead to the emergence of organizational strategies, conscious or not" (Vaara & Whittington, 2012). A conceptual framework for research on strategizing, building on Aristotle, involves explicit, purposeful strategy processes and non-deliberate, purposive and emergent strategizing through everyday practices (Chia & Holt, 2006; Chia & Rasche, 2010). Explicit strategizing is a deliberate action, and is guided by predefined goals, implying that strategizing is an act of purposeful intervention (Chia & Rasche, 2010). Explicit strategizing is understood through Aristotle's ideas of *episteme* and *techné*. *Episteme* involves context-independent, explicit and propositional knowledge (Chia & Rasche, 2010), while *techné* refers to codifiable techniques and practical instructions (Chia & Rasche, 2010). Emergent strategizing is non-deliberate, responding to situations, and viewed as purposive, practical coping (Chia & Rasche, 2010). Emergent strategizing is found through *phronesis* and *mētis*. *Phronesis* is practical wisdom "acquired through experience that accounts for the ability to perform expediently and appropriately in defined social circumstances" (Chia & Rasche, 2010). *Mētis* is the "practical intelligence required to escape puzzling and ambiguous situations" and the "ability to attain a surprising reversal of unfavorable situations to achieve favorable outcomes" (Chia & Rasche, 2010). Both explicit and emergent strategizing in MNCs are highly relevant to understand how strategies are operationalized so that firms can benefit from open innovation. The four different categories of open innovation, together with the conceptual framework of strategizing, are helpful when looking at the different MNC strategizing activities for innovation. This applies for both the HQ and the periphery, as well as for the Norwegian and foreign-based MNCs in Norway.

3. Research Design and Method

To analyze how explicit and everyday coping practices are conducted in relation to innovation in MNCs, we had to uncover recurrent practices in several MNCs. We performed a collective case study where a number of cases were chosen purposefully to investigate the topic under investigation (Stake, 1994). To explore strategizing in Norwegian and foreign-based MNCs, firms within the oil and gas service sector and in a specific region, were targeted. Ten MNCs were selected. We assumed that they use different types of innovation and strategizing to remain competitive, both locally and globally. Table 2 lists the key characteristics of the sample firms.

3.1. *Data collection*

To inquire about both local and global innovation activities, as well as related strategizing, we needed to talk with people who understood both local and global activities. They also needed to

Table 2. Characteristics of the case firms.

Firms — anonymized	HQ location	Number of employees globally
Alpha	Corporate HQ: USA	80 000
Beta	Corporate HQ: USA	70 000
Gamma	Corporate HQ: USA	65 000
Delta	Corporate HQ: USA	126 000
Epsilon	Corporate HQ: Europe	58 000
Zeta	Corporate HQ: Europe	14 000
Eta	Corporate HQ: Norway	2200
Theta	Corporate HQ: Norway	800
Iota	Corporate HQ: Norway	28 000
Kappa	Corporate HQ: Norway	600

have strategic roles, including the focus on the innovative activities performed within the company. Hence, we conducted in-depth interviews with CEOs, country managers, chairpersons and/or business developers within the firms. Strong efforts were made to obtain interviews with these top managers, and therefore we conducted one series of interviews in all 10 firms in autumn 2013 and another in 2014. The 2013 interviews facilitated a further set of interviews with other top managers within the firms. The main method of data collection was in-depth interviews, with a semi-structured interview guide. Between two and four persons were interviewed in each firm. Several top managers insisted on having a business developer or business unit manager present during the interview, resulting in a total of 20 conducted interviews. Each interview lasted between one and two and half hours, and were recorded and later transcribed.

3.2. *Data analysis*

The data analyses went through several phases. After the first interviews, we wrote descriptions of each case company. This was then approved by the top managers of the firms to validate the veracity of the empirical material, facilitate feedback, and enhance the trustworthiness of the analysis (Lincoln, Y. S. & Guba, 1985). We also developed coding frameworks for emerging themes, using NVivo qualitative data analysis software for the coding analyses. The emerging themes were defined as the different types of innovation activities, and the answers were categorized in relation to explicit or emergent strategizing. After the second series of interviews, all the empirical material was scrutinized according to what was conducted locally, if a subsidiary, and centrally if it was a HQ, in addition to other localities. The main difference between the two series of interviews was our awareness of the different innovation activities in relation to locations. The results were different depending on whether the HQ had its location in Norway or if the Norwegian location was a subsidiary and hence part of the periphery. The views were almost opposing, they were consistent but from two different viewpoints: HQs focused on purposeful strategies whereas the periphery focused more on the tasks they had to follow from HQ, while also accentuating their coping practices for everyday work and thus purposive strategizing.

In line with the mystery construction of Alvesson and Kärreman (2007), our analyses progressed by going back and forth between in-depth analyses of the empirical material and the theoretical framework. During the process, we encountered two puzzles. The first puzzle was how explicit strategizing of innovation openness by the acquiring companies, including having subsidiaries at dispersed locations, also led to innovation and technology within the companies. This finding made us examine the recurring practices of "company acquisitions" and "location sourcing" within and among the cases in more detail. The second puzzle was how the coping and everyday practices of performing activities, the emergent strategizing, led to innovation. Local specialization was used to acquire activities within the global firm, at the same time as activities were channeled to the knowledgeable person, regardless of location. The acquisition of activities was empirically interesting because it often went against explicit strategizing or focus from HQ. Theoretically, "activities acquisition" was of interest because we could not find adequate theoretical concepts for the phenomenon (Alvesson & Kärreman, 2007). "Company acquisition" as a phenomenon is broadly understood and theorized, while the acquisition of activities seems less considered.

"Knowledge sourcing" was puzzling because the knowledgeable persons to whom the activities were channeled were not necessarily physically located at the related business unit. Several top managers explained how key employees live in different places across the globe, more in relation to where they come from than to the unit to which they belong. "Knowledge sourcing" was found regardless of whether these employees were part of the top management team, innovation activities or business development. We did not find adequate theoretical explanations for this phenomenon, apart from common-sense understanding, and chose to use this as the last construct. The two puzzles: (1) how explicit strategizing for openness enhanced intra-firm innovation, and (2) how emergent strategizing and everyday practices caused dispersed performance of activities and future innovation, led us to further theorizing in relation to ambidexterity, and to exploitation and exploration activities, as well as push and pull forces in intra-bound innovation. This is our theoretical contribution.

4. Findings

The 10 firms operate in a highly competitive global market. Market demands are high in relation to performance, quality, reliability and standards compliance, also with a strong focus on cost efficiency. Demanding customers drive innovation both locally and globally, forcing firms to remain competitive. In the following, we first focus on the explicit corporate HQ strategy activities in relation to innovation, both "company acquisitions" and "location sourcing". We then focus on the emergent strategies and everyday practices from the subunits in relation to innovation, exposing "activities acquisition" and "knowledge sourcing".

4.1. *Acquiring companies*

Across the cases, we found that acquiring companies locally is a clear HQ strategy to be innovative. As a top manager in Beta explained:

> *"I think it is more a very deliberate strategy to buy up innovative companies, including Norwegian, and then globalize the technology ... We have innovation at many levels within the company, but we also buy up technology. And it is clear, we have grown to 63,000 employees during the last 17 years or something from 1000. So, it is clear that it does not happen by organic growth. It's, we've acquired an enormous number of companies."*

A manager in Alpha explains: "We have an active policy of buying start-ups". A Zeta manager emphasized that "Zeta is itself a result of a merger in 2010 between two large companies with different owners". Acquiring innovative companies is understood as a normal way of innovating, as explained by the chairman of Delta:

> *"All the large innovations are acquired innovations. Delta has brought on and further developed all of them. The organization is on a constant look out for companies to acquire and has had a lot of success with this strategy so far. Delta has been very good at acquiring companies without simply swallowing them up; the companies may still continue development and manufacturing, often even under their original names."*

For foreign-owned firms, there is an explicit strategy to acquire Norwegian companies. Epsilon Norway, for instance, has acquired several Norwegian companies. An Epsilon manager explained:

> *"Norway is a source for new technology, and one of Epsilon Norway's functions is to pay attention to new technological developments that happens in the market. The industry has a certain dynamic, and it is a good sign if an actor can succeed in the demanding market that Norway is with regard to regulations and standards for quality."*

Not only is Norway an attractive market to acquire companies from, but the MNCs use their global network to commercialize from innovation, as a manager in Alpha explained:

"So, we've wanted things to happen in Norway, and we bought Norwegian technology companies such as A and B. There is a lot of good Norwegian technology. ... when it comes to Company A, that is a good example since we, after the acquisition, had work in Norway, and when Alpha got their hands-on Company A, we went global with the technology fairly quickly. In comparison, Company A would have taken many years of work to get it distributed globally, while Alpha has a network."

A manager in Gamma described its focus in acquisitions:

"We are always looking for companies to buy.... We do buy small companies every year. They complement us. They are not competitors. They have small overlapping areas. But they were bought specifically to complement us. We integrate the people as well. Generally, we buy the people and the companies, although we do buy IP as well, occasionally, of course."

Another way is to acquire certain percentages of companies. The manager for Gamma explained:

"... They've had a cash injection from us. Other large players are also a part of them [innovative small company] in fact. And we will see what they do. We might try and increase our interest in them totally or we might just continue as they are. And with the smaller companies of course, and this is a smaller company, I don't know how many people, but it is less than 15. If a large company buys them, they tend to swallow them and they just get lost. Where if you let them continue to do what they do best, which is to innovate and focus, then they might grow into something more interesting before we sell them. This happens occasionally. You know, as a global enterprise we do a lot of things. But generally — this goes

for most companies — you would tend to take a majority ownership in a company."

When reflecting on Norway as a particularly interesting location for acquiring companies, the answers were negative, as a manager in Alpha explained:

"I do not believe that Norway is either under- or overrepresented in relation to either 1: company or 2: technology acquisitions. When it comes to corporate acquisitions it is generally being profiled and appears in the media, in a country like Norway, which has a slightly open media philosophy. Technology acquisitions do not reach the media and we read very little about those."

The practice of acquiring companies is based on explicit strategizing from the HQ to obtain growth, incorporate technical expertise and acquire innovation and technology. The acquisitions are often integrated into the MNC, where innovation and technology are spread globally internally in the company. How innovation is spread globally within the company, however, depends on its location sourcing, which is exposed next.

4.2. *Sourcing locations*

The HQs and HQ functions are often spread out in different regions, hence reflecting a work division within the MNCs. Location sourcing refers to the formal organizing structures that are linked to the division of activities between the various locations. Two Beta managers reflect on the local Norwegian position in relation to their HQ in the US. The HQ representative is placed in Norway, whereas the rest of the HQ team has its location in Houston. He explained:

"The different locations are pretty specifically chosen. Each location exists because they are a bit unique and it has its specific expertise, and access to the expertise, not least. So, around

Norway, everything that goes on in mechanical design is in that location. This includes the headquarters, management, and management of large projects."

A manager at Gamma described its sourcing locations as follows:

"Norway is part of a super region, we call it. We have a headquarters in London to run Europe, Africa, Russia, and the Caspian. Our super region completions director sits in Norway, and so does our super region drilling director. He sits in Norway as well. So, he reports to people in Aberdeen. And because they don't need to sit in London, they are on a plane most of the time. They have to be here or in Russia or the Caspian or Angola or Nigeria or whatever. But they have team members in other places as well. He has a team of 10 or so support people, a few of which are Norwegian ... the driver for this has been to become more multicultural and closer culturally to the customer, as well as geographically.

An Iota manager explained:

"We have a more flat organizational structure, with closeness to locations where the activities are performed."

There were large differences between the case firms depending on where they had their multinational corporate HQ, whether it was in Norway or in other parts of the world, mostly in the US. These differences were related to the focus. Those with HQs in Norway emphasized how the offices located around the world could be best supported financially and administratively by HQ. One example is Eta, having 30 offices worldwide and employing over 70 nationalities. The CEO explained the location sourcing as follows:

"We have an international division of labor. We have a global market and regional centers; Houston covers South America with offices in Rio and Canada who report to Houston, so the UK and

Singapore with offices reporting in from Perth and others. We have regional hubs and in each regional hub, we have 4 business units globally. One is designated as coordinator for the region, and the 4 business areas with 4-line managers. We have some features that are only located here in Norway, regarding economic and administrative tasks. We are distributed in technical disciplines; an expert can sit in Australia working against the whole system. It is more a question of where the best people are located. In Houston, we have very many talented people and in London, as the other two major offices ... this [Oslo] is not the biggest office. We have experts in disciplines in other locations, for instance the processing of data occurs as much in Houston as in Norway...."

The Theta CEO from the Norwegian HQ explained the rationale behind the location sourcing, by focusing on the reduction of the economic administrative burden in the local offices:

"What we do, is to liberate our local managers with what they call economic administrative stuff. When they render top service to our customers, then everything that is economic and administrative should be taken away. We have to relieve them, and we become an administrative hub, so they can be heroes and take care of our customers. Everything we can centralize, we centralize here. What we shall not centralize is the local creativity. Documentation is here, logistics and purchasing, personnel and HR are centralized."

Those having HQs in other countries, and where the top managers were interviewed locally, emphasized their locality's role in relation to other locations. A manager in Alpha explained:

"Geographically, Scandinavia is a very big business for Alpha. That turnover and size of the organization is far beyond the major operations that we have in North America, so the Scandinavia Operation is one of the largest internationally both in sales and in the number of employees. So, the status of these facilities here and the people who work here, is relatively high abroad."

Several locations also have a local HQ function, as a business developer in Alpha explained:

> *"headquarter function to Alpha in Scandinavia is located in Stavanger. So, it means that we do not have a global or regional function. We have only one locality. And 'local' means Norway and Denmark."*

The analyses revealed differences in focus between the HQs and the subunits. The findings are, however, that these two different views complement each other. Some of the findings are that local innovation or technology needs are found locally, the technology is thereafter developed centrally, and then it is tested and used locally again before global distribution. A manager from Alpha explained how the technology development happens centrally at HQ:

> *"The company has the technology development centrally in general. The assignments for technology come from local needs, i.e., the technology needs that arise locally will provide work for development taking place centrally at HQ.... all the demands which stem from the North Sea, we have successfully developed technology within the company which has first been used on the Norwegian shelf and been distributed globally afterwards."*

From this, we argue that there are both clear strategic goals and work divisions behind the location sourcing within the MNCs. It seems as if local needs drive innovation, while most of the official R&D takes place at the HQ location centrally. Through location sourcing, the MNCs are close to the innovation drivers locally, and several of the HQ activities are dispersed locally. Dispersed HQ activities are either performed locally because top managers are physically located outside of the HQ, as in Beta, or performed regionally through regional HQs. Location sourcing within MNCs is also a well -defined and explicit strategy to be close to the customers. However, there are many innovation activities performed locally that are not explicitly defined, which we define as "acquiring activities".

4.3. *Acquiring activities*

We found a functional division of different types of activities within the MNCs, and labeled this division "activities acquisition". Activities acquisition refers to how the innovation activities within MNCs are acquired. A Zeta manager explained how innovation activities are part of everyday work:

> *"Some projects are standard projects, give or take, and others are more demanding. There tends to be a degree of operational innovation in all projects. Innovation occur almost every day, during operations. In general, the development regards how to do things better, safer and more efficiently, or how to solve problems that arise because of new demands, harsher weather conditions, deeper sea and so on."*

There are many innovation activities within the MNCs that are locally bound, even though the official HQs are located elsewhere, having large R&D and innovation centers centrally. An Epsilon manager explained the struggle between HQ and local innovation activities as follows:

> *"There is a force from the head office wanting to attach much of the development closer to the operating head office in Houston. This means that Epsilon Norway constantly has to justify the use of resources for development in Norway."*

Regarding innovation developed locally, the Eta manager explained:

> *"We started from scratch. We developed it internally in Eta ... This has been a revolution within the industry. With this technology, we add to our prices ... Our competitors try, but they do not manage to match our quality. Within this segment, we are market leaders globally."*

Delta in Norway is a miniature of Delta globally, in that it has activities related to all parts in the value chain. In addition to this, all research in this field in Delta is performed in Norway.

"R&D in Norway, compared to revenue in Norway, is nearly twice as large as global R&D compared to global revenue. About 10% of the global budget for R&D is spent in Norway. In total, about 4000 employees in Delta Norway work with R&D."

explained the Chair of the Board. A Beta manager explained activities acquisition as:

"In such large projects ... consists of 20 people in Stavanger, and so triple in another location in Norway. This is about engineering, also we have project managers in a separate area. Physically we deliver this network here.... Stavanger becomes a kind of integration point for everything in the project globally."

Activities acquisition is not understood as an explicit strategy, but rather part of the everyday operations that develop over time. The practices are related to performing value-creating activities, including innovation and how innovation activities are increasingly performed locally. Hence, activities acquisition are activities that are less clearly defined, and do not need the same amount of justification from HQ of time and resources used. The outcome, however, is that activities from other locations are acquired and performed locally. Furthermore, activities acquisition leads to innovation through customer projects and is later spread internally and globally in the respective companies Activities acquisition is closely related to knowledge sourcing.

4.4. *Sourcing knowledge*

Knowledge sourcing was also found to be highly important in the periphery. Knowledge sourcing refers to how know-how is distributed throughout the company. The CEO of Eta explained:

"Regarding technical expertise, it is all distributed in relation to where the clever people are. It all depends on the people. In Houston, we have many excellent people, and in London. Those are the two other big offices apart from Oslo."

A Gamma manager explains knowledge sourcing from a local perspective with effects globally:

"So, we deliver real-time drilling services. We support them from Stavanger. It is actually a process that has been enabled by technology, by real-time broadband communications. We do occasionally support rigs in Nigeria and other places in the world. So, we have exported that competence globally from here ... a good number of that global group are Norwegians ... They are travelling and setting up Operation Support Centers and are helping with processes in Brazil or Russia or whatever. — [The Norwegian know-how is] Diffused. And of course, in Norway it is very special. There are not many places in the world that have a broadband capability to more or less all of their rigs offshore. It's very unusual. But in Norway, that is how it is. And that is the enabling technology. It is not our technology. But we've piggy backed on it. So, I can literally sit here, and I can tell a drill strength up in the Barents Sea to turn left, go down, go right ... whatever. And then, 20 seconds later: Ohh, wrong way. Yes <laughter>. So, it's an amazing capability ... I don't know how many centers we've got globally, but we've got lots of them and they've been primarily set up by Norwegian led know-how expertise."

Zeta has engineers all over the world, but there are three main engineering centers: Scotland, France, and Norway. One manager explained:

"Norway is one of these offices with strong technological capability and expertise especially in this segment, while Scotland and France represent other competencies. In addition to this, in the UK we have extensive knowledge on a nearby field. All in all, the organization has organically strong centers of expertise around the world."

The Iota top manager explained:

"We would have saved money if locating in the UK, but in Norway we have a core competence that is dependent on a close cooperation with the places where our activity is conducted."

Knowledge sourcing is different from location sourcing because it is not an explicit strategy, but more part of an emergent strategy. Knowledge sourcing is related to where the competent people are, but also to the expertise they develop over time in relation to different activities performed. Knowledge sourcing is therefore part of emergent strategy.

5. Analysis

Analyzing the patterns of explicit HQ strategy activities like: (1) company acquisitions and (2) location sourcing, and the emergent strategies of (3) activities acquisition and (4) knowledge sourcing, we uncovered the dynamic relationship and related strategizing between the various patterns We also examine whether these four patterns are exploitative or explorative and whether the patterns can be related to *episteme*, *techné*, *phronesis* or *mẽtis* (Chia & Rasche, 2010). To view the mechanisms, we also looked into which pull and push forces that are at play. We discuss these four patterns in relation to strategy and innovation.

Company acquisition is how innovative firms are acquired by, and used within, the MNC. There is a clear strategy from the HQ to take over, acquire and internalize external expertise and knowledge through company acquisitions, helping the acquired companies to commercialize through MNCs' global distribution channels. This finding is well aligned with the open innovation literature. Firms acquire expertise from outside through inbound pecuniary innovation (Dahlander & Gann, 2010). Firms even increase their innovative output through acquisitions (Ahuja & Katila, 2001). Hence, through such company acquisition, the innovations and expertise may be revealed externally as in outbound non-pecuniary innovation (Dahlander & Gann, 2010). Company acquisitions are exploitative because the goal of the MNC is to access new markets and complement existing services to enhance a company's technological position. Company acquisitions expose an explicit strategy of pull through buying external know-how, technology and innovation, and integrating this into the MNC. Once integrated into the MNC, however, the innovation/technology is pushed out and commercialized through the global intra-firm network to be adapted in local contexts (outbound open innovation strategy).

Location sourcing is about the division of activities at different localities. From the HQ, there is a clear strategy of establishing subsidiaries at different locations to be closer to the drivers of innovation (demanding customers and needs), closer to specific expertise, closer culturally (technically and environmentally) and closer geographically. However, the HQ requests that official R&D takes place at the central HQ location. This exposes inbound non-pecuniary innovation and how external sources of innovation are sourced into the companies through local presence (Dahlander & Gann, 2010). This finding is also well aligned with international business literature; firms exploit their innovations through location sourcing (Patel, 1995). Location sourcing is explorative because the companies slowly build up their innovation activities locally, having to defend the use of resources to HQ. Location sourcing pushes the MNC to different locations due to the need for proximity to the drivers of innovation. In this study, the localized engineering-based hubs in the oil and gas sector globally and their potential customers and suppliers. There is also a pull back to HQ for specific knowledge, and especially for R&D purposes.

Activities acquisition is about how innovation activities within MNCs are acquired at subsidiaries on the periphery. The operational innovation occurring every day in subsidiary locations becomes an integration point for whole projects. HQ works to carry out the innovation and R&D centrally, while the subsidiaries justify their innovation activities locally. Activities acquisition is not the same as inbound pecuniary innovation since it is more in-house, driven from operational, projects and customer needs. We thus call this intra-bound pecuniary open innovation. Activities acquisition is explorative because the different actors innovate from daily work and the activities gradually evolve. The periphery exploits that knowledge, or know-how, which is distributed by assuring that activities are being performed where the knowledgeable employees are located. Activities acquisition exposes pull through everyday coping activities locally, and push to get more of the same or related activities from the company globally. This is an ongoing interactive process of push and pull, and can be an important enabler for innovation in the MNC, and, as such, it activates both local and global sources of knowledge.

Knowledge sourcing is about how knowledge is dispersed through the company, independent from location. Activities are thus channeled to a knowledgeable person with expertise within a field. At the same time, these professionals are building strong centers of expertise all over the world. Intensive within-firm collaboration helps geographically distant subsidiaries to reach higher levels of technological effectiveness (Liefner *et al.*, 2013). Knowledge sourcing is not the same as location sourcing, since the sources are internally in the firms. Thus, we call this intra-bound non-pecuniary innovation. Knowledge sourcing first exposes push to use those knowledgeable people regardless of where they are located, and then pull to strengthen centers of expertise all over the world, thereby strengthening the knowledgeable experts for the firm globally.

From our empirical material, we found that company acquisition and location sourcing are explicit strategizing activities from HQ. Through everyday practices on the periphery, we find that activities acquisition and knowledge sourcing are patterns of coping practices and strategizing performed at subsidiaries. Explicit strategizing thus stems from the HQ, while emergent strategizing activities are performed locally (Figure 1).

This understanding of explicit vs. emergent strategizing only highlights part of the picture. *Company acquisition*, how firms are acquired by and used within the MNC, is exploitative, following *episteme*, is context independent and explicit (Chia & Rasche, 2010). *Location sourcing* is explorative through dividing activities, following *techné,* referring to codifiable techniques and practical

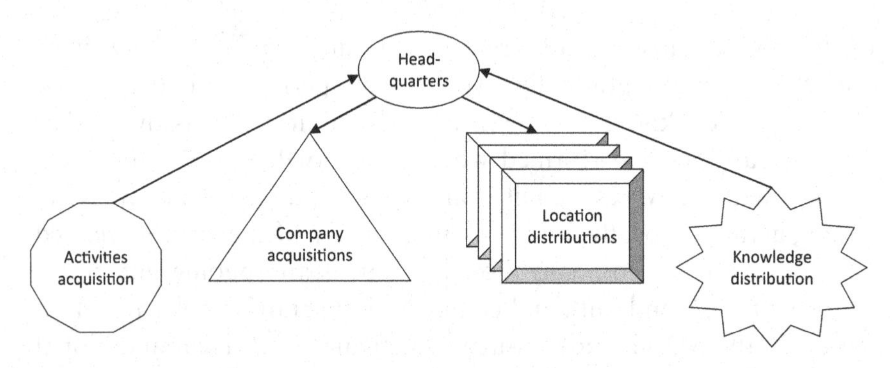

Figure 1. Center and periphery strategizing.

Table 3. The geography of the firms' strategizing practices for innovation.

	Exploitation	Exploration
Explicit *How*	*Company acquisition — how innovative firms are acquired by and used within the MNC* *Episteme*	*Location sourcing — how activities are divided* *Techné*
Why	Push by helping the acquired companies to commercialize through the global distribution channels of MNCs. Pull by buying up and internalizing external expertise and knowledge through company acquisitions	Push to different locations: closer to the drivers of innovation (demanding customers and needs), closer to specific expertise, closer culturally (technically and environmentally) and closer geographically Pull back to HQ for R&D
Emergent *How*	*Knowledge sourcing — how knowledge is distributed through the company* *Phronesis*	*Activities acquisition — how innovation activities within MNC are acquired* *Mētis*
Why	Push follows knowledge and expertize Pull building strong centers of expertise all over the world	Push: operational innovation occurring every day, subsidiary location becomes an integration point for whole project Pull: HQ works to carry out more R&D in HQ in a constant struggle

instructions. *Knowledge sourcing,* the way knowledge is distributed through the company, is exploitative and follows *phronesis*, the ability to perform appropriately, acquired through experience (Chia & Rasche, 2010). Finally, *activities acquisitions*, how innovation activities are acquired, is explorative, following *mētis*, being the "practical intelligence required to escape puzzling and ambiguous situations" (Chia & Rasche, 2010). The findings are schematically shown in Table 3.

6. Implications and Conclusions

We set out to answer the question: *how* and *why* do strategizing activities occur in MNCs in relation to open innovation? To answer our research question, regarding *how* strategizing activities occur, we found explicit strategizing from HQ, exposing exploitative focus through company acquisitions and explorative focus through location sourcing. The strategizing on the periphery included everyday coping practices showing explorative focus through activities acquisition and exploitative focus through knowledge sourcing.

Our contribution nuances and contradicts Regnér's (2003) findings. Regnér argues that strategizing at the periphery is understood as inductive, while strategizing at the center is found to be deductive. Our findings, however, suggest that there is ambidexterity both at the center and at the periphery. This implies that the HQ's role in strategizing is to facilitate exploration and exploitation while being receptive to different strategies of exploration and exploitation from the periphery. Furthermore, our findings suggest that explicit strategizing from HQ has an exploitative focus through company acquisitions, but an explorative focus through location sourcing. Emergent strategizing is explorative through activities acquisitions but exploitative through knowledge sourcing. We thus found that there is parallel center ambidexterity and periphery ambidexterity. Also, strategizing is centralized and dispersed in parallel. The performance of ambidexterity exposes how the different explicit and emergent strategizing practices are enacted.

To answer *why* strategizing occurs in relation to innovation, we examined a variety of push and pull forces from HQ and subsidiaries. These forces also differ according to exploitation and exploration strategies, as well as whether there is an explicit or emergent firm strategy. However, they all have elements of innovation. Inbound innovation is found through explicit strategies, such as company acquisitions and location sourcing, whereas innovation-related activities are pulled both to HQ and from HQ to the subsidiaries. The emergent inductive processes of knowledge sourcing and activities acquisition show the forces of both push and pull in intra-bound innovation. Explorative innovation activities evolve through everyday activities of problem solving. This underpins why the activities are acquired by those who know or are channeled to the knowledgeable person. These strong push forces (from HQ to subsidiary location) might be typical in learning processes that are tacit and customer-specific by nature, such as in services. Knowledge sourcing represents *phronesis*, whereas activities acquisition represents *mētis*.

These findings expand Dahlander and Ganns (2010) four categories of open innovation. They expose inbound and outbound innovation to interactions that are pecuniary vs. non-pecuniary. Our findings, expose that inbound innovation should be further nuanced with intra-bound innovation, highlighting acquiring activities as pecuniary and sourcing knowledge as non-pecuniary innovation (Table 4). We also nuanced open innovation theorizing, emphasizing both deliberate and purposeful practices and emergent and purposive practices.

In the study on how the MNCs strategized for open innovation mechanisms, we did not expose outbound innovation. Future

Table 4. Nuancing Dahlander and Gann (2010) different forms of openness with intra-bound innovation.

	Inbound innovation	**Intra-bound innovation**
Pecuniary	Acquiring companies	Acquiring activities
Non-pecuniary	Sourcing locations	Sourcing knowledge

research should explore whether and how there are different forms of outbound innovation such as extrabound innovation, further building on our findings of intra-bound innovation.

A limitation of this research is not being able to uncover the strategizing through a practice-based study over time. Such a study would focus on the actions and activities forming part of each of the four identified practices. However, this could be a theme for future research. Also, future research should investigate whether explicit and deliberate strategizing stemming from the HQ, is related to radical innovation, but also whether emergent and non-deliberate strategizing activities, that are locally performed, are more related to incremental innovation.

A focus on explicit and emerging strategizing, as well as *episteme*, *techné*, *phronesis*, and *mẽtis*, has been viewed together with ambidexterity and open innovation. We contribute to the ambidexterity literature by demonstrating parallel center and periphery ambidexterity and that strategizing is both centralized and dispersed in parallel. We also show that the role of HQs in MNCs is not the only driving force. Furthermore, we contribute to open innovation theorizing by introducing *intra-bound innovation*. This is an addition to the existing understanding of inbound and outbound innovation.

This chapter shows how explicit global strategizing from HQ spur internal open innovation dynamics and emergent practices locally on the periphery. These subsidiaries continuously define and redefine innovation intra-bound through practical coping in customer projects and, as such, strengthen MNC positions globally. The empirical material shows how the push and pull forces in the center and periphery form part of the interactive processes of open innovation internally in firms.

Acknowledgments

We wish to thank Heidi Wiig for her advice over various stages of this project. This work was supported by the Research Council of Norway (233737).

References

Ahuja, G., & Katila, R. (2001). Technological acquisitions and the innovation performance of acquiring firms: A longitudinal study. *Strategic Management Journal*, 22(3), 197–220.

Alvesson, M., & Kärreman, D. (2007). Constructing mystery: Empirical matters in theory development. *Academy of Management Review*, 32(4), 1265–1281.

Bianchi, M., Cavaliere, A., Chiaroni, D., Frattini, F., & Chiesa, V. (2011). Organisational modes for open innovation in the bio-pharmaceutical industry: An exploratory analysis. *Technovation*, 31(1), 22–33.

Chesbrough, H., & Crowther, A. K. (2006). Beyond high tech: Early adopters of open innovation in other industries. *R&D Management*, 36(3), 229–236.

Chesbrough, H. (2017). The future of open innovation. *Research-Technology Management*, 60(1), 35–38.

Chesbrough, H. W. (2003). *Open innovation: The New Imperative for Creating and Profiting From Technology*. Harvard Business School Press, Boston.

Chesbrough, H. W. (2012). Open innovation: Where we've been and where we're going. *Research-Technology Management*, 55(4), 20–27.

Chia, R., & Holt, R. (2006). Strategy as practical coping: A Heideggerian perspective. *Organization Studies*, 27(5), 635–655.

Chia, R., & Rasche, A. (2010). Building and dwelling world-views — Two alternatives for researching strategy as practice. In: D. Golsorkhi, S. David, L. Rouleau, and E. Vaara (eds.), *Cambridge Handbook of Strategy as Practice*. Cambridge University Press, Cambridge, pp. 34–46.

Chia, R., &MacKay, B. (2007) Post-processual challenges for the emerging strategy-as-practice perspective: Discovering strategy in the logic of practice. *Human Relations*, 60(1), 217–242.

Dahlander, L., & Gann, D. M. (2010). How open is innovation? *Research Policy*, 39(6), 699–709.

Hydle, K. M. (2015). Temporal and spatial dimensions of strategizing. *Organization Studies*, 36(5), 643–663.

Hydle, K. M., Aas, T. H., & Breunig, K. J. (2014). Strategies for financial service innovation: Innovation becomes strategy-making. In: A.-L. Mention and M. Torkkeli (eds.), *Innovation in Financial Services: A Dual Ambiguity*. Cambridge Scholars Publishing, Cambridge.

Jarzabkowski, P., & Spee, A. P. (2009). Strategy-as-practice: A review and future directions for the field. *International Journal of Management Reviews*, 11(1), 69–95.

Jarzabkowski, P., Balogun, J., & Seidl, D. (2007). Strategizing: The challenges of a practice perspective. *Human Relations*, 60(1), 5–27.

Kornberger, M., & Clegg, S. (2011). Strategy as performative practice: The case of Sydney 2030. *Strategic Organization*, 9(2), 136–162.

Liefner, I., Wei, Y. D., & Zeng, G. (2013). The innovativeness and heterogeneity of foreign invested hightech companies in Shanghai. *Growth and Change*, 44, 522–549.

Lincoln, Y. S., & Guba, E. G. (1985). *Naturalistic inquiry*. Beverly Hills, Calif: Sage Publications.

Mintzberg, H., & Waters, J. A. (1985). Of strategies, deliberate and emergent. *Strategic Management Journal*, 6(3), 257–272.

Patel, P. (1995). Localised production of technology for global markets. *Cambridge Journal of Economics*, 19(1), 141–153.

Regnér, P. (2003). Strategy creation in the periphery: Inductive versus deductive strategy making. *Journal of Management Studies*, 40(1), 57–82.

Stake, R. E. (1994). *Case Studies*. Sage Publications, London.

Tsoukas, H. (2010). Practice, strategy making and intentionality: A Heideggerian onto-epistemology for strategy-as-practice. In: D. Golsorkhi, L. Rouleau, D. Seidl, and E. Vaara (eds.), *The Cambridge Handbook of Strategy as Practice*. Cambridge University Press, Cambridge, pp. 47–62.

Vaara, E., & Whittington, R. (2012). Strategy-as-practice: Taking social practices seriously. *The Academy of Management Annals*, 6(1), 285–336.

https://doi.org/10.1142/9789811234491_0007

Chapter 7

New Service Development Process: What Can We Learn from Research and Technology Organisations?

Pierre-Jean Barlatier[*,§], *Eleni Giannopoulou*[†,¶] *and Lidia Gryszkiewicz*[‡,||]

[*]*EDHEC Business School 393, Promenade des Anglais — BP 3116 06202, Nice Cedex 3, France*

[†]*Université de Strasbourg, CNRS, BETA UMR 7522, F-67000 Strasbourg, France*

[‡]*The Impact Lab sàrl 29, Boulevard Grande-Duchesse Charlotte, L-1331, Luxembourg*

[§]*pierre-jean.barlatier@edhec.edu*

[¶]*elina.giannopoulou@gmail.com*

[||]*lidia.gryszkiewicz@theimpactlab.org*

Abstract. Research and Technology Organisations (RTOs) are crucial contributors to national innovation systems, as they provide highly innovative services to European economies and societies. However, very few empirical studies have investigated how RTOs actually develop their service innovations. Thus, this paper explores the new service development (NSD) process of five renowned European RTOs. Our findings suggest that the NSD process in RTOs is ad hoc and highly iterative, and mostly informal. Moreover, this research highlights two additional noteworthy results: the NSD project vs. service innovation "culture" dilemma and the interesting open dimension of the RTOs' NSD process.

Keywords. Service innovation; Research and Technology Organisations; New Service Development Process.

1. Introduction

The rise of the tertiary sector has turned service innovation into a major business concern for companies striving to gain a competitive advantage over their rivals. However, service innovation is also relevant for institutions other than companies. In Europe, innovation is largely facilitated by actors known as Research and Technology Organisations (RTOs). These are mostly large, dedicated institutions that focus on service innovation for the benefit of the economy, society and/or environment, actively contributing to national innovation systems. Therefore, RTOs are confronted with service innovation challenges. Yet, few empirical studies have addressed specific and practical issues of service innovation — such as the new service development (NSD) process — within RTOs, as most of the studies on RTOs focused on a macro level, mainly dealing with their impact on their environment (Arnold *et al.*, 2007; Barge-Gil & Modrego-Rico, 2008). The purpose of this chapter is to study how RTOs deal with the development of new services to meet the expectations of governments, companies and society as a whole.

To do so, we followed a methodology of qualitative, explorative, multiple case studies (Miles & Huberman, 1994; Yin, 2003). We investigated five European RTOs through multiple semi-structured interviews and secondary data analysis. As a result, we identified the approaches that these RTOs use towards their NSD process, as well as typical dilemmas and issues they face in this regard. Our insights are beneficial for understanding the mechanisms at work at RTOs, how they contribute to the service innovation debate in general, and how they could serve for other organizations dealing with NSD challenges.

In this chapter, we first introduce our theoretical background, followed by our empirical research method. Subsequently, we present our findings regarding the five studied RTOs. In the discussion, we provide a critical analysis of our findings. Last are conclusions and perspectives for further research.

2. Theoretical Framework

2.1. *Services, service innovation and their specificities*

Services are currently a major economic activity for developed countries. The special nature of services makes it challenging to provide a unique and commonly accepted definition of what a service is. That is why services are often described on the basis of their special characteristics that differentiate them from goods: intangibility, simultaneity, variability (de Jong *et al.*, 2003; de Brentani, 1991).

Moreover, the difficulty in defining and thoroughly understanding services has led to the misleading assumption that services are not innovative or are merely peripheral to industrial/product innovations (Gadrey *et al.*, 1995). As a result, the literature on service innovation has been more fragmented and less empirically grounded than the academic developments regarding innovation process in the manufacturing sector (Toivonen, 2011; Gadrey *et al.*, 1995).

However, this is not up to par with the importance of service innovation in the current European economic context, in which more than 70% of EU total value added originates from service industries (EUROSTAT, 2011). RTOs represent a specific type of organization that is at the core of European service innovation. In the next chapter we explain, in more details, what RTOs are and why they are relevant to this research.

2.2. *Defining RTOs*

According to the (European Association of Research and Technology Organizations (EARTO), RTOs are "organizations which as their predominant activity provide research and development, technology and innovation services to enterprises, governments and other clients ..." (EURAB, 2005). They are positioned between academia and industry, having overlapping activities with the two. Moreover, they have strong links with the government, as they qualify for public funding (Arnold *et al.*, 2010). RTOs' mission is to have impact on industry by helping companies (especially SMEs) move "one step beyond" their existing capabilities and reduce the risks associated with innovation, for a faster rate of economic development (Arnold *et al.*, 2007). Therefore, while European RTOs' turn over amounts to approximately EUR 18.5–23 billion, their actual economic impact is considered much higher, in the range of EUR 25–40 billion (Arnold *et al.*, 2010). Moreover, RTOs also have strong societal and environmental impacts in terms of service innovation.

The operating models of RTOs have raised interest over the years and can be generally defined as comprising the following stages: (i) exploratory applied research in order to develop an area of capability or a technology platform often in collaboration with academia, (ii) further refinement, often in collaboration with industry, and (iii) exploitation of this knowledge via consulting, licensing and spin-off company creation (Arnold *et al.*, 2007, 2010). The next section deals with how RTOs actually develop their new services.

2.3. *The new service development process in RTOs*

Although RTOs are clearly important service innovative organizations, we could not find evidence of any previously developed model specifically addressing their NSD process. As a result, we turned to the relevant general service innovation literature in order to identify NSD models that could serve as a framework for our research.

Originally, the NSD-related academic work was largely based on new product development (NPD) models (de Jong *et al.*, 2003; Kline & Rosenberg, 1986; Cooper *et al.*, 1994; Saren, 1984). But the validity of NPD models has not been demonstrated in the context of services (Stevens & Dimitriadis, 2005). Several attempts have been made to model the NSD process separately. First, models structuring the NSD process were created in a linear way, including the 10-step model of Shostack (1984), the 15-stage model of Scheuing and Johnson (1989) and the 16-stage one by Edgett and Jones (1991), to name a few. The weaknesses of these linear models for NSD were quickly identified, as they did not reflect the intertwined, unstructured and cyclical reality of the NSD process. Thus, scholars moved to the iterative way of presenting NSD. Bitran and Pedrosa (1998) created a model consisting of six stages; namely strategic assessment, concept development, system design, component design, implementation, feedback and learning. Tax and Stuart (1997) created another iterative representation of NSD which shows that new services can be born in or of existing service systems. Furthermore, Johnson *et al.* (2000) developed a four-stage model with 13 internal tasks. De Jong *et al.* (2000) see the NSD process as a process comprising only two stages — the search stage and the implementation stage — with different activities in each one. Finally, Froehle and Roth (2007) proposed their non-sequential resource-based conception of NSD in which they emphasize the interdependence of design and development phases in NSD. In general, the iterative models are perceived as better describing the NSD process and its intangible and often abstract nature.

For the purposes of our research, we thus needed a model that is both iterative and flexible, and generic enough to be adapted to the specific context of RTOs. The model best meeting these requirements is that of Johnson *et al.* (2000). We believe this model, which is described in details below, is flexible and iterative enough to reflect the particular nature of service innovation, and generic enough to cover the basic aspects of NSD while allowing consideration of RTOs' characteristics. Moreover, it has already been successfully used in an empirical study (Froehle & Roth, 2007).

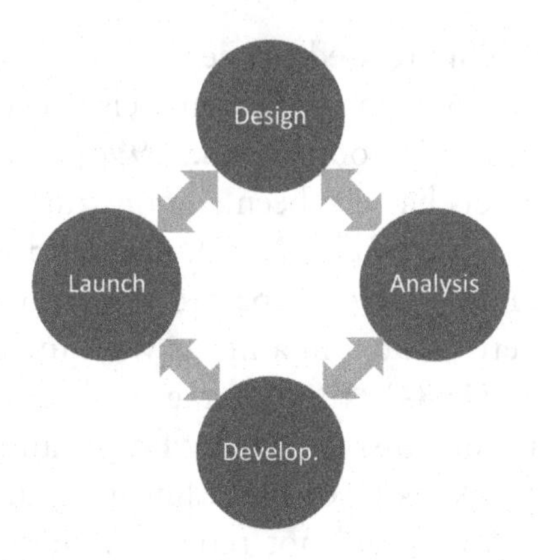

Figure 1. The four-step generic model of NSD (adapted from Johnson *et al.*, 2000).

The NSD model of Johnson *et al.* (2000) comprises four steps (Figure 1). The *design stage* includes the generation and screening of the new service idea, the formulation of the related objectives, as well as the concept development and testing. In the *analysis stage* decisions are made about the project's authorization or stopping. The *development stage* comprises the actual service design, as well as the marketing plan and piloting. Finally, in the *launch phase,* all the full-scale launch activities to the customer, as well as post launch review, are performed. These steps can take place in different sequences, allowing for several cycles of iterations, before the actual service is developed.

3. Method

Due to the novel nature of our research objectives, we followed a research methodology based on qualitative, explorative, multiple case studies (Yin, 2003; Miles & Huberman, 1994). Our sample consisted of five European RTOs which possess certain maturity in service innovation but present differences in size, location, organizational setting and target markets. The studied RTOs were selected in collaboration with the EARTO. Through a set of semi-structured

RTO	Short description	Interviewees
RTO A	RTO A is a European organisation that has about two thousand employees with various research focus areas such as materials, energy, information and communication and electronics, among others. RTO A has recently put strategic focus on services. The organisational design of this research centre is carefully deployed with an aim to address and balance its different functions; namely R&D, contract and applied research.	Research director Marketing director Technology manager Quality manager Development manager Research scientist/project managers (2) Professor
RTO B	RTO B has a very good reputation in the country in which it operates. It employs thousands of researchers working in very different knowledge domains, working in different locations, including some foreign offices. Different parts of the organisation focus on different topics and cooperate strongly with different universities. As RTO B is a very large and diverse organisation, we focused our research on one of its departments only, the one most concerned with service innovation.	Competence / business line senior manager (2) Project manager (1) Researcher (3) Marketing manager (1)
RTO C	RTO C employs several thousand people, recently reorganized along research topics and expertise areas. The organisation has also recently introduced a new strategy, in which the main focus is to increase impact (on both the industry and society) through demand-driven research. Besides the research activities, RTO C also focuses on incubating and spinning off new companies through an associated business structure. There are multiple locations in the country and some satellite offices abroad.	Strategy team (3) Business line manager / researcher (4) Scientific research coordinator (1) Competence manager (1)
RTO D	RTO D employs more than four hundred people and aims to offer science-based services in various fields, such as health, materials, IT and management among others. This RTO has also recently strategically shifted towards services. In this study we focus on an IT and management service-dedicated department of more than one hundred employees.	Director of Department Service Line manager (2) Program manager R&D manager Service Portfolio manager R&D engineer (2) Unit manager Quality manager
RTO E	RTO E has about two thousand employees. It focuses on the ICT field and is a major European player in ICT research to propose innovative ICT services and usages. RTO E is specialised in education and training as well as technology transfer and incubation. It has several locations and subsidiaries in the country.	Innovation Director Innovation & Development manager (2) International relationship coordinator SME project manager SME Club manager Media & Usages Project manager

Figure 2. Overview of case RTOs and interviewees.

interviews with key service innovation actors in each RTO (see Figure 2) and the complementary analysis of secondary data, we gathered data on the different approaches that these RTOs use to develop and deliver new services.

In total, we conducted 41 interviews (lasting 75 min on average); all were recorded and transcribed. Interviewees were first asked about their role and discussions progressively focused on NSD-related activities. Given our exploratory research targets, we also encouraged interviewees to give their opinion freely in order to better identify best practices, as well as related issues. Transcribed interviews were communicated to interviewees for checking, validation and, if necessary, complementary information. Collected data were analyzed according to a coding grid derived from our theoretical framework, with additional emerging codes progressively added throughout the process, leading to grid stabilization. Data analysis was supported by software-facilitated iterative double-coding. Disagreements were resolved by discussion to reach an inter-coder agreement of more than 90% for each code, confidently above the threshold recommended by Miles and Huberman (1994). Finally, single-case analyses were followed by cross-case analysis techniques (Miles and Huberman, 1994, Yin, 2003), leading to the findings presented in Section 4.

4. Findings

None of the cases studied had a formalized NSD process. Although the different elements of Johnson's (2000) NSD model existed in reality (in varying levels of maturity) they were not formally recognized as such, nor did they follow any fixed sequence. In general, three types of NSD behaviors could be distinguished: (1) a "reactive" response to an expressed client need — such as development of a new Virtual Reality lab as an answer to a client's need to test the user friendliness of their new facility before the construction stage would even begin, (2) a "proactive" internal investment in the development of new service with no particular client contract signed yet, such as investment in developing a service as a response to an emerging observed megatrend, and (3) a (reactive or proactive) combination or extension of existing services, e.g., development of lighting concepts for working environments — based on existing services related to visual technologies, ergonomics in the workplace and improve energy efficiency.

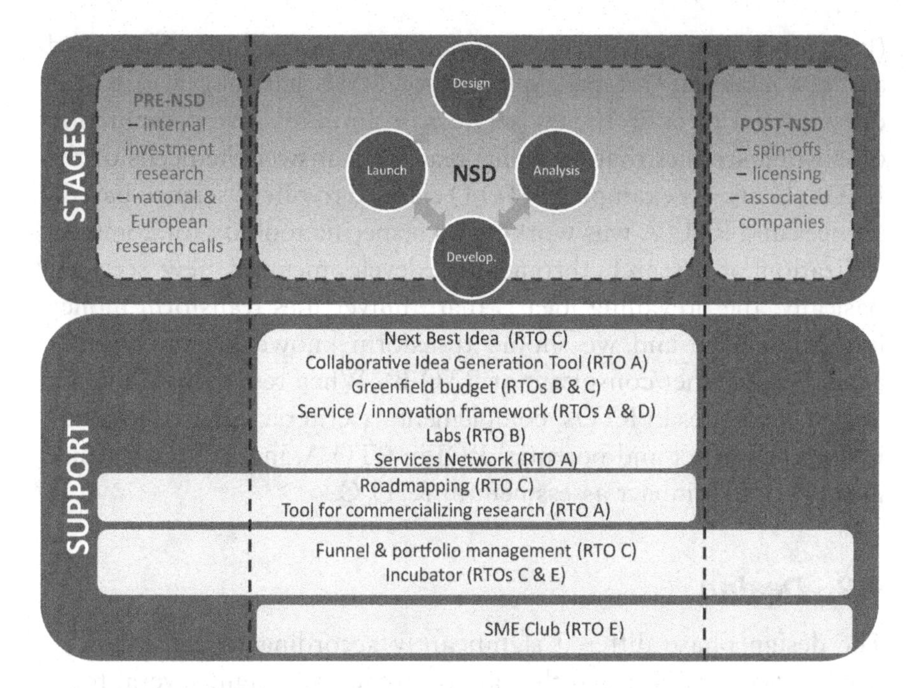

Figure 3. NSD in the studied RTOs: Stages and the relevant support.

Moreover, different "stages" were often blurred, intertwined, parallel or absent, depending on the type of service provided (i.e., the level of service standardization, type of service content, service sector etc.). Interestingly, most interviewees did not see any need to change this situation, stating that "having one process doesn't solve the problem" (Top Manager 2 RTO A). However, while there was no formal NSD procedure, various tools to support the NSD process were available across our sample, including creativity initiatives, management frameworks and tools, or even equipment. We discuss the related activities in the following sections, as depicted in Figure 3.

4.1. *Pre-NSD research activities*

The services offered by the RTOs are strongly linked to research activities; in fact, "there is no innovation without research"

(Innovation Director, RTO E). Inputs from research are often used as "raw material" or prerequisites for NSD that may lead to the development of entirely new services or an innovative combination of existing services that may also reactively answer the needs of the RTOs clients. As examples, RTO D claimed to offer "science-based" services and RTO A was working on a specific toolbox for commercialization of research through the development of new services. Basically, the prevailing idea is that "universities transform money into knowledge and we should transform knowledge into money again" (researcher/consultant at RTO B). When research is inherent to NSD activities in RTOs, complementary criteria are used to guide it, such as impact and potential ROI in RTO A, industrial relevance in RTO B and impact assessment in RTO C.

4.2. *Design*

The design phase differed significantly according to proactive vs. reactive types of NSD. In the case of proactive design, several tools could be found in the different RTOs. RTO A had a software-based collaborative platform for idea generation. RTO B was constantly searching for new trends by being proactively involved in various market interactions, fairs, conferences and workshops. RTO C organized a "next best idea" contest to stimulate organizational creativity, and used formal "road-mapping" extensively to decide on the most prospective trends and technologies. At RTO D, the ideation phase was enhanced through a dedicated organizational project involving the development of an organization-wide collaborative creativity platform. Moreover, technological, scientific and market watch served as sources for new ideas. Generally, though, this stage of NSD was characterized by minimal formalization in the RTOs studied. Openness towards creativity ("there is always the opportunity to come up with new ideas", Research Topic Manager, RTO C) and easily allocated start-up money were important elements.

In reactive service development the procedure was more abstract and the use of specific tools was minimal, as the ideation phase depended on collaboration with clients, in which problems and solutions were defined in a more *ad-hoc* way. In this kind of NSD project, the active involvement of the responsible client managers, such as those in RTOs A and D, was essential.

4.3. *Analysis*

As mentioned earlier, the stages of the NSD were not at all clear-cut across our sample. Still, based on extensive coding, we were able to distill some insights concerning each group of activities we could put under the umbrella "steps" labeled 'analysis, development, launch and post-NSD. These are further presented below.

For the majority of the studied RTOs, analysis was "the most important step in the NSD process in terms of working intensity" (Unit Manager, RTO D). Various tools with predefined criteria for the evaluation of new service ideas were used, with all the RTOs applying some kind of formal screening method:

> *"You also have to stay realistic. I mean, you have one brain and two hands. You cannot follow all your crazy ideas you have. And the important question is, which, how to choose and that is difficult" (Competence Centre Leader, RTO B).*

However, the analysis step was by far most advanced at RTO C. The organization used a detailed funnel management approach, including a stage-gate system, in order to constantly evaluate the state of the NSD projects in its portfolio. Each project was assessed according to a strict set of pre-defined criteria including market attractiveness, business attractiveness, potential impact, level of innovativeness, etc. Only after meeting those criteria was a service moved through a gate to reach the next development stage.

Note also that for RTOs A and C the identification and stopping of unpromising project ideas was not perceived as taboo, or was even encouraged, as at RTO C:

> *"To avoid people are afraid that they get a bad (evaluation] (...) we said 'when you have the guts to stop it, it is a positive sign not a negative sign'" (Top Manager — RTO C).*

4.4. *Development*

The development stage is crucial, as "innovation — this is 10% idea and 90% processing" (Innovation Director, RTO E). However, contrary to the analysis stage, we found no evidence of any formalized procedure in the development stage of the NSD process. Moreover, it was often adapted to meet specific client needs in a better way, as one researcher at RTO B put it, "we propose a solution but we adapt it to the customer needs (...) we are normally very happy to adapt all the processes". Across all the cases studied, it was in the development stage that client involvement was most intensive, in fact, there was "a constant collaboration with the customer" (Middle manager, RTO A).

Interestingly, only RTO B had specific facilities to aid service development, such as labs or 3D visualization rooms.

> *"(...) I think what we realized is that we have to come up with ways to show and prove what we're doing and make it more interesting because we don't have a product to sell (...) because it's pretty boring to just hand people over a pile of papers (...)" (Marketing & Communication Manager, RTO B).*

Generally, it was acknowledged that the less innovative the service, the easier it was to apply a more standard procedure to its development. For instance, in all the RTOs studied, service tests were merely performed for more "tangible" (e.g., software-related) services.

4.5. *Launch*

The launch phase of a new service could happen either at the beginning or at the very end of the NSD process, according to the nature of the service. In the case of reactive services, the launch phase could be blurred with the other stages and could start at the beginning of the NSD, as the collaboration with the client was immediate and constant. In the case of proactive services, contact with the clients started later as the RTOs communicated the launch of a new service with brochures, public presentations, or training to attract potential clients.

Every RTO had a client satisfaction feedback system, but it usually focused on the end result rather than on the NSD process itself. Sometimes, however, reflections at the end of the process served as a trigger to "transfer the approach of that project to other ones as well" (Researcher, RTO B).

Nevertheless, in general, the launch phase was "done on a non-structured and relatively informal way" (Researcher, RTO B), constituting the least fixed part of the NSD process in most cases.

4.6. *Post-NSD*

Once the NSD process was finished, there were many ways of transferring the service to the market. New services could lead to the creation of spin-offs, licensing, or even consulting, but the main point was that the service was externalized once it was acknowledged as being mature (RTO D and E) or when deploying it became routine work, from which researchers could no longer learn (RTO C). This externalization was done either towards external actors, or towards an associated company dedicated to the commercial exploitation of designed services (RTOs A and C), or it could even take the form of a new start-up (RTO E). In the latter case, RTO E had even created a "SMEs Club" to continue to support its own spin-offs, as well as to interact with some other SMEs they collaborated with.

4.7. *Beyond the NSD process*

According to our theoretical framework, several aspects that go beyond the NSD process itself, but are strictly related to it, require our attention: the NSD strategy, the project-service dilemma and the inclusion of stakeholders in the process.

4.7.1. *NSD strategy*

Generally, RTOs' mission stressed the need for positively influencing the economy, society and environment through their NSD efforts.

However, the importance of this "impact" dimension varied among the RTOs. For instance, in RTO C this was crucial and specific impact teams had been created to operationalize the concept, while in RTO B the notion of broader impact was not too prevalent, due to the greater focus on the direct client.

Another point is related to the strategic focus on services. In RTOs A and D, services were explicitly mentioned in the strategy in the past few years and relevant actions to support them had recently been undertaken (new organizational service-related initiatives, rethinking of existing departments towards service development, organization of service symposiums, creation of service-related positions, etc.). In contrast, RTOs B and C seemed more experienced in working with services and apparently did not feel the need to explicitly focus on a specific service approach in their strategy.

4.7.2. *NSD project vs. service*

There is a thin line between the concepts of "project" and "service" in project-based service organizations such as RTOs. A service can be either a one-off unique assignment or an umbrella of multiple similar projects. The degree to which different RTOs distinguished the two differed. For example, RTO A has started a service network bringing people from the whole organization together as an attempt to put more emphasis on the service, without abandoning the project setup. RTO C, in its funnel and portfolio management approach, defined product-market combinations (which could be anything from a single project to an elaborate long-term set of projects) as the level of decision-making. RTO B, interestingly, did not really keep itself busy with the discussion between services and projects.

Different levels of organization within a single RTO often had a different focus. Especially in RTO D, we identified a gap between top management-induced service focus and project-based work-floor reality. Specifically, while the top management was developing high-level strategic service frameworks, in practice, the only framework the employees were following, in their NSD project, was strict quality management requirements.

4.7.3. *The inclusion of various stakeholders in the NSD process*

The NSD was influenced significantly by several stakeholders in all our cases. The most important one was the client, as all RTOs aimed to work closely with the client organizations throughout the NSD process. Clients were seen as sources for new ideas in all RTOs and client potential was taken into account in the evaluation of new project proposals. Of course, clients were especially strongly involved in the client-driven type of service innovation, where the collaboration often spanned throughout the whole NSD process.

Other stakeholders, such as partners from academia, associations, and public authorities, were also seen as important contributors to the NSD process. More specifically, in RTO A there were initiatives to include external partners in the internal services network; in RTO C large consortia and knowledge arenas were often developed (especially in the development "step" of the NSD), while RTO D was concerned with detailed stakeholder analysis, when proactively developing new services (typically mostly, but not exclusively, during the "analysis" step). Finally, and specifically as far as collaboration with academia was concerned, some challenges were identified in RTO A regarding the different objectives of the collaborative parties. In RTO B, in contrast, as the links with academia were very tight, no such issues were expressed. Overall, collaboration with academia tended to be more intensive in the first "steps" of the NSD, mostly analysis, and was most prevalent in the "proactive" type of service innovation.

5. Discussion

5.1. *RTO's NSD processes*

The "blurring" of the stages of NSD activities in RTOs empirically confirms the intertwining of stages in the NSD model of Johnson *et al.* (2000). Interactions between the design and analysis stages were particularly strong. Moreover, two additional NSD stages proved to be important in this specific organizational context, which we labeled pre-NSD and post-NSD. This result is congruent with

RTO innovation models that include research and knowledge/technology transfer (as presented by Arnold *et al.*, 2010). Hence we propose to add these two stages in an NSD process dedicated to RTOs because (i) both pre- and post-NSD are strongly linked with other "classical" NSD steps to propose a complete NSD process; and (ii) they can be subdivided into common, generic practices, and supported by dedicated resources, i.e., defined as genuine organizational capabilities.

Another important finding regarding NSD processes in RTOs was the identification of what we labeled "proactive services" and "reactive services". Proactive services are new services that are RTO-led, coming from an internal service idea, e.g., R&D driven, while reactive services are new services that come from external service sources, e.g., a demand from a client. This echoes Toivonen and Tuominen's (2009) identification of different processes leading to an innovation in knowledge-intensive business services, with on the one hand (i) the R&D model (i.e., R&D driven) and (ii) rapid application (i.e., test of a new service idea that is quickly brought to markets and developed) as "proactive" NSD, and on the other hand (iii) the practice-driven model (i.e., a service co-developed with the client including possibilities for replication), which corresponds to a "reactive" behavior from the RTO. Hence, our results confirm Toivonen and Tuominen's (2009) model, and especially in the case of RTOs, we believe that different generic NSD processes (with different succession of stages) can be associated to Toivonen and Tuominen's service types. For instance, in the case of a rapid application, the launch phase of the NSD starts immediately, while in the case of R&D-driven service, this step happens at the end of the NSD. Further research could examine RTOs' NSD processes according to these three distinct types in order to try to identify generic processes.

5.2. *The project-service dilemma*

As we have seen, different RTOs deal differently with the potential distinction between a project and a service. Some, in fact, seemed to exhibit a degree of "service obsession", trying very hard to "develop a service culture" (RTO A) or "make a shift from projects to services" (RTO D). While innovation and project development are sometimes perceived as "an impossible equation" (Aggeri & Segrestin, 2007), we could argue that project focus is a natural setup for non-standardized services and RTOs' financing model and that this situation does not *per se* exclude having a broader vision. In fact, collaborative paths developed in projects support both exploitation of existing capabilities and exploration of new ones (Manning & Sydow, 2011).

Moreover, projects' entrenchment in more permanent contexts allows them to draw from this context's resources (Sydow *et al.*, 2004). Of course, to fully reap the benefits of a project setup, organizations should consciously manage their ability to transfer project-based learning to the organizational level, traversing the "learning boundaries" (Scarbrough *et al.*, 2004). The same, however, is true for extracting experiences from existing services for the benefit of NSD. Den Hertog *et al.* (2010) label that as service innovation capabilities of (un-)bundling, scaling and stretching. Therefore, irrespectively of the "project" vs. "service" discussion, we suggest that organizations such as RTOs could focus on providing better support for transforming on-off experiences into re-usable service solutions, e.g., through deployment of a more proactive quality management process, implementation of an advanced knowledge management system, or granting the researchers additional time for combining and transferring their experiences.

5.3. *The open dimension of the NSD process in RTOs*

The inclusion of several stakeholders in the NSD process was evident in all the RTOs studied. This was expected, as RTOs are the interface between industry and academia, and have special links with the government (Arnold *et al.*, 2007, 2010). They need to manage and often balance different parties and incentives in the innovation process (big consortia of partners, academia collaboration, clients involvement), which was confirmed by our findings.

However, we argue that the innovation process in RTOs goes a step further than usual cooperation, namely that it bears many of the characteristics of what Chesbrough (2003) coined the "open innovation" paradigm. Moreover, RTOs are active in both dimensions of open innovation (Chesbrough *et al.*, 2006): the inward and the outward. RTOs are not merely concerned with assimilating external knowledge by including external actors in their own innovation process (e.g., through customer inclusion or big panels of partners in our cases), but they are also concerned with external commercialization activities, as technology transfer and/or start up creation is a special element of their mission and innovation model (Arnold *et al.*, 2010; Berger & Hofer, 2010); the most representative case here being RTO E. Finally, note also that the case of services makes it even more relevant for RTOs to work in an open way. As services represent a unique experience for each client and they are tailored to specific needs, they call for a more open and collaborative approach towards innovation, where the inclusion of the client throughout the NSD process can be very beneficial (Chesbrough, 2011). This is in line with Perks *et al.* (2012) who argue that if there is radical service innovation this is very likely to be created through co-creation.

Thus, we can conclude that the RTOs can very rarely, if ever, adopt a closed innovation model for the NSD process because their special characteristics "force" them to innovate openly. The special nature of RTOs and their position in the innovation systems, being placed between academia and industry implies a role of an active collaborator and intermediary in innovation (Agogué *et al.* 2013). As a result, they have to develop the open innovation capabilities to allow them to collaborate with several partners in the services innovation process. This special element makes them very interesting, as the lessons that can be learned from these organizations can be beneficial for all kinds of service organizations struggling with the challenges of openness in innovation.

6. Conclusions and Suggestions for Further Research

This study regarding RTOs' NSD processes highlights several points of interest. First, although the RTOs investigated had some NSD steps of the NSD process in place due to their focus and their experience in service innovation, none of them acknowledges having a formal NSD process, but rather tools to support it. This is in line with the finding of Crevani *et al.* (2011) that "researchers stress a need for formalized processes for development work, while practitioners focus on facilitating innovation in everyday operations" (p. 177). This also confirms and strengthens the *ad hoc* and empirical "embeddedness" of the NSD stages as presented in Johnson's *et al.* (2000) model. Moreover, the specificities of service innovation in an RTO context call for adding two phases to Johnson *et al.*'s NSD process. We defined these steps as pre-NSD and post-NSD stages, which illustrate respectively the importance of research and science and the RTO's mission of value transfer towards market and society (i.e., the notion of "impact"), beyond the sole delivery of a new service to a client.

Second, a noteworthy result concerns the organizational culture and systems beyond the NSD process of the RTO itself. Actually, we detected issues at project-based RTOs that have quite recently chosen to strategically focus on services and are experiencing difficulties reconciling a project-based organization and its inherent systems with a service culture. In most cases this has consequently led to some organizational design changes (e.g., the service network) that still need to mature.

Third, we noticed that the NSD activities of RTOs may involve several external actors and contributors, such as clients of course, but also government, academia, other RTOs, etc., revealing the increased importance of openness in the case of RTOs.

Many research paths are pointed out throughout this article. The most evident is related to the definition of RTOs' specific NSD processes according to each type of service innovation. Also, further investigation about the "project-service" dilemma, in RTOs, may provide knowledge about how such organizations can articulate project-based organizational systems with a corporate, shared "service" culture, regarding quality, knowledge management, etc. Another important contribution could involve studying the specific position and role of RTOs within an open service innovation system, in order to explore their contributions and interactions with their environment and stakeholders.

References

Aggeri, F., & Segrestin, B. (2007). Innovation and project development: An impossible equation? Lessons from an innovative automobile project development, *R&D Management*, 37, 37–47.

Agogué, M., Yström, A., & Le Masson, P. (2013). Rethinking the Role of Intermediaries as an Architect of Collective Exploration and Creation of Knowledge in Open Innovation. *International Journal of Innovation Management*, 17(2), 1–24.

Arnold, E., Brown, N., Eriksson, A., Jansson, T., Muscio, A., Nählinder, J., & Zaman, R. (2007). The role of industrial research institutes in the national innovation systems. *Analysis VA 2007:12*. VINNOVA.

Arnold, E., Clark, J., & Javorka, Z. (2010). Impacts of European RTOs. A study of social and economic impacts of research and technology organizations. *A report to EARTO*. Technopolis group.

Barge-Gil, A., & Modrego-Rico, A. (2008). Are technology institutes a satisfactory tool for public intervention in the area of technology? A neoclassical and evolutionary evaluation. *Environment and Planning C: Government and Policy*, 26(4), 808–823.

Berger, M., & Hofer, R. (2010). *The Internationalization of R&D: How about Research and Technology Organizations? Some Conceptual Notions and Qualitative Insights from European RTOs in China*. Joanneum Research Institut für Technologie und Regionalpolitik, InTeRegs, Wien.

Bitran, G., & Pedrosa, L. (1998). A structured product development perspective for service operations, *European Management Journal*, 16, 169–189.

Chesbrough, H. (2003). *The New Imperative for Creating and Profiting from Technology*. Harvard Business Publishing, Boston, Massachusetts.

Chesbrough, H. (2011). *Open Services Innovation: Rethinking Your Business to Grow and Compete in a New Era*. Jossey Bass, San Francisco.

Chesbrough, H., Vanhaverbeke, W., & West, J. (2006). *Open Innovation: Researching a New Paradigm* Oxford University Press, Oxford.

Cooper, R. G., Easingwood, C. J., Edgett, S., Kleinschmidt, E. J., & Storey, C. (1994). What distinguishes the top performing new products in financial services? *Journal of Product Innovation Management*, 11, 281–299.

Crevani, L., Palm, K., & Schilling, A. (2011). Innovation management in service firms: A research agenda. *Service Business*, 5, 177–193.

de Brentani, U. (1991). Success factors in developing new business services. *European Journal of Marketing*, 25, 33–59.

de Jong, J. P. J., Bruins, A., Dolfsma, W., & Meijaard, J. (2003). Innovation in service firms explored: What, how and why? *Literature Review. Strategic Study B200205*, EIM Business and Policy Research.

den Hertog, P., van der Aa, W., & de Jong, M. W. (2010). Capabilities for managing service innovation: Towards a conceptual framework, *Journal of Service Management*, 21, 490–514.

Edgett, S., & Jones, S. (1991). New product development in the financial service industry: A case study. *Journal of Marketing Management*, 7, 271–284.

EURAB. (2005). *Research and Technology Organizations (RTOs) and ERA* [Online]. Brussels. Available: https://www.earto.eu/wp-content/uploads/2005_12_01_EURAB_RTOs_and_ERA.pdf [Accessed 02/2021].

EUROSTAT (2011). *European Economic Statistics — 2010 Edition*. 2010 ed. Luxembourg.

Froehle, C. M., & Roth, A. V. (2007). A resource-process framework of new service development. *Production and Operations Management*, 16, 169–188.

Gadrey, J., Gallouj, F., & Weinstein, O. (1995). New modes of innovation — How services benefit industry. *International Journal of Service Industry Management*, 6, 4–16.

Johnson, S. P., Menor, L. J., Roth, A. V., & Chase, R. B. (2000). *A Critical Evaluation of the New Service Development Process*, Sage, Thousand Oaks, CA.

Kline, S., & Rosenberg, N. (1986). *An Overview of Innovation*, National Academy Press, Washington.

Manning, S., & Sydow, J. (2011). Projects, paths, and practices: Sustaining and leveraging project-based relationships. *Industrial and Corporate Change*, 20, 1369–402.

Miles, M. B., & Huberman, A. M. (1994). *Qualitative Data Analysis*, Thousand Oaks, Sage.

Perks, H., Gruber, T., & Edvardsson, B. (2012). Co-creation in radical service innovation: A systematic analysis of microlevel processes. *Journal of Product Innovation Management*, 29(6), 935–951.

Saren, M. A. (1984). A classification and review of models of the intra-firm innovation process. *R&D management*, 14, 11–24.

Scarbrough, H., Swan, J., Laurent, S., Bresnen, M., Edelman, L., & Newell, S. (2004). Project-based learning and the role of learning boundaries. *Organization Studies*, 25, 1579–1600.

Scheuing, E. E., & Johnson, E. M. (1989). New product development and management in financial institutions. *International Journal of Bank Marketing*, 7, 17–21.
Shostack, G. L. (1984). *Service Design in the Operating Environment*, C American Marketing Association, Hicago, IL.
Stevens, E., & Dimitriadis, S. (2005). Managing the new service development process: Towards a systemic model. *European Journal of Marketing*, 39, 175–198.
Sydow, J., Lindkvist, L., & DeFillippi, R. (2004). Project-based organizations, embeddedness and repositories of knowledge: Editorial. *Organization Studies*, 25, 1475–1489.
Tax, S., & Stuart, F. I. (1997). Designing and implementing new services: The challenges of integrating service systems. *Journal of Retailing*, 7, 58–77.
Toivonen, M. (2011). Different types of innovation processes in services and their organizational implications. In: G. Faiz and F. Djellal (eds.), *The Handbook of Innovation and Services — A Multidisciplinary Approach*, Edward Elgar Publishing Limited, Cheltenham, UK, pp. 221–249.
Toivonen, M., & Tuominen, T. (2009). Emergence of innovations in services. *The Service Industries Journal*, 29, 887–902.
Yin, R. K. (2003). *Case Study Research: Design and Methods* (3rd edn.). Sage Publishing, California.

https://doi.org/10.1142/9789811234491_0008

Chapter 8

Outbound Open Innovation in Tourism: Lessons from an Innovation Project in Norway

Tor Helge Aas[*,†,§], *Kirsti M. Hjemdahl*[†,¶], *Daniel Nordgård*[*,†,||] *and Erik Wästlund*[‡,**]

[*]*School of Business and Law, University of Agder, Campus Kristiansand, Universitetsveien 25, 4630 Kristiansand, Norway*

[†]*NORCE Norwegian Research Centre, Nygårdsgaten 112, 5008 Bergen, Norway*

[‡]*Karlstad University, Universitetsgatan 2, 651 88 Karlstad, Sweden*

[§]*tor.h.aas@uia.no*

[¶]*kihj@norceresearch.no*

[||]*daniel.nordgard@uia.no*

[**]*erik.wastlund@kau.se*

Abstract. The concept of open innovation was introduced by Henry Chesbrough in 2003 and refers to firms' use of inflows and outflows of knowledge to improve innovation processes. The concept has received considerable scholarly attention, but most research has focused on how manufacturing firms can manage inflows of knowledge during their product innovation processes. How outflows of knowledge can be managed by service firms during their innovation processes has until now not received the same scholarly attention. In this chapter, we therefore aim to contribute in filling this knowledge gap by observing an innovation project in tourism during its implementation. The findings suggested that tourism firms reveal different types of knowledge to other tourism firms in non-pecuniary outbound open innovation processes. In this case the knowledge was revealed in joint workshops where several firms participated and in bi-lateral meetings between two firms, and sometimes the knowledge was transferred via consultants or researchers that acted as "knowledge mediators". The findings also suggested that tourism firms decided to reveal knowledge to other firms to improve their image, increase the market size and to sharpen up their own business. Implications for management as well as the need for further research are discussed in the chapter.

Keywords. Innovation in services; open innovation; tourism management.

1. Introduction

The concept of open innovation was introduced by Chesbrough (2003) and has received considerable scholarly attention in subsequent years (Chesbrough & Bogers, 2014). The concept has been defined as "the use of purposive inflows and outflows of knowledge to accelerate internal innovation, and to expand the markets for external use of innovation, respectively" (Chesbrough *et al.*, 2006, p. 1). In this definition two fundamentally different types of open innovation may be identified. The first type, often called inbound open innovation (e.g., Dahlander & Gann, 2010), is perhaps the most intuitive type and refers to the case where a firm gets or buys access to knowledge from an actor external to the firm, and uses this knowledge to enable, improve or accelerate its internal innovation processes. The second type is perhaps less intuitive, but nevertheless of great value in many situations, and refers to the case where a firm sells or reveals its knowledge to an external actor that in turn use this knowledge to enable, improve or accelerate its innovation processes.

The second type is often called outbound open innovation (e.g., Dahlander & Gann, 2010) and is the focus of this chapter.

Results of research on outbound open innovation processes have suggested that selling or out-licensing so-called "spill-over" knowledge from internal R&D activities may increase firms' return of R&D investments (e.g., Nerkar, 2007) and it has also been suggested that revealing knowledge without charge to external actors will sometimes be beneficial for firms in the long run because this behavior may induce collaboration (Henkel, 2006). Despite these, and other, valuable findings, however, our knowledge and understanding of outbound open innovation remains limited (West & Bogers, 2017). In fact, while a considerable number of studies have investigated why and how firms implement inbound open innovation processes, the investigation of why and how firms implement outbound innovation processes have been addressed by considerably fewer studies (West & Bogers, 2014).

Another fact is that the majority of open innovation studies have used empirical data from product innovation processes in manufacturing (Huizingh, 2011). At the same time the results of general service innovation research have indicated that there are considerable differences between service innovation processes and product innovation processes in manufacturing. Knowledge resulting from traditional R&D activities are for example in general found to be less relevant for service innovation processes than for product innovation processes (Droege *et al.*, 2009). As also indicated by Chesbrough (2010), it seems likely that this, and other characteristics of service innovation, may affect the relevance of outbound open innovation in service firms, as well as how outbound open innovation should be implemented in this context. However, the lack of empirical studies focusing on outbound open innovation in service firms implies that our knowledge in this area is particularly limited.

Due to the fact that the service industry consists of a number of sub-sectors that are heterogenous to some degree, both with regard to the characteristics of the value propositions they offer and with regard to the innovation processes they implement (Srholec & Verspagen, 2012), we have chosen to focus on one particular sub-sector that is particularly different from manufacturing, namely tourism, in this chapter. On that account we empirically explore the following research questions (RQs) in this chapter:

- **RQ1:** How is knowledge that flows out from tourism firms during innovation processes transferred?
- **RQ2:** What types of knowledge flow out from tourism firms during innovation processes?
- **RQ3:** Why does, or does not, tourism firms decide to reveal or sell knowledge to other firms?

The chapter is structured in the following manner: In the next section we discuss theoretically how the characteristics of innovation in services may affect the relevance and implementation of outbound open innovation practices in this context. In the section thereafter we present the qualitative method chosen to address the research questions. The empirical findings are provided in the fourth section and in the fifth section we discuss these findings and conclude.

2. Theory

Dahlander and Gann (2010) suggest that we can distinguish between two different types of outbound open innovation processes: Pecuniary and non-pecuniary. Pecuniary processes refer to processes where "firms commercialize their inventions and technologies through selling or licensing out resources" (Dahlander & Gann, 2010, p. 704). Through this type of outbound open innovation processes firms get payed for inventions that they do not want to commercialize through their own business model. Thus, pecuniary outbound open innovation processes may be understood as a way to fund R&D activities that the business managers find irrelevant for the renewal or replacement of the firm's value propositions or business model. According to research it has become more common for firms with internal R&D departments to implement R&D strategies where the out-licensing of knowledge, for example in the form of patents, is targeted (e.g., Fosfuri, 2006). Research has documented that this strategy may be successful for firms, especially in knowledge-intensive sectors such as the pharmaceutical industry (e.g., Hu *et al.*, 2015). Nevertheless, considerable challenges are also associated with this R&D strategy. These challenges are for example related to the high level of transaction costs that often occur when patents or other types of intellectual property are sold (e.g., Gambardella *et al.*, 2007) and the general difficulty in valuing these types of assets (Chesbrough & Rosenbloom, 2002).

Non-pecuniary open innovation processes refer to processes where *firms reveal internal resources without immediate financial rewards* (Dahlander & Gann, 2010, p. 703). The value of this type of open innovation is hard to measure and has been called both mysterious and attractive by prior authors (e.g., Wang & Zhou, 2010). The disadvantages of this type of open innovation are obvious: When a firm reveals its internal resources, for example in the form of knowledge, to an external actor, there is a risk that the firm will not be able to capture any benefits (Helfat, 2006). Nevertheless, research has indicated that although this practice does not have any immediate financial rewards for the revealing firm, the revealing firm may in many cases enjoy some long-term indirect benefits. The findings of Wang and Zhou (2010), for example, suggested that the revealing of internal knowledge enabled inbound open innovation processes for firms that in turn influenced innovation performance in a positive manner. Furthermore, the findings of other authors have suggested that when firms in an industry share knowledge and technology it may lead to numerous innovation initiatives that in turn may be beneficial for most firms that belong to the industry (Allen, 1983).

Despite these important insights, there are still a number of knowledge gaps related to how firms in practice implement outbound open innovation practices. In fact, a recent literature review by West and Bogers (2017) states: *While there is considerable research that systematically investigates inbound OI [open innovation], our understanding of outbound OI remains heavily influenced by anecdotal examples of Chesbrough (2003)* (p. 44). This literature gap is especially true for innovation in services. Most studies discussing outbound open innovation focus on product innovation in manufacturing (West & Bogers, 2017), and due to differences between product innovation and service innovation (Droege *et al.*, 2009) it is uncertain whether the current knowledge about open innovation in general, and outbound open innovation in particular, may be transferable to a service context (Aas & Pedersen, 2016), such as tourism.

Research on innovation in tourism has suggested that *tourism enterprises rarely have R&D departments or other dedicated resources for innovation* (Hjalager, 2010, p. 5). Instead *search and knowledge acquisition processes take place in a more complex and informal manner* in tourism (Hjalager, 2010, p. 5) and tacit knowledge is found to play an important role during innovation processes in tourism firms (Cooper, 2006), as well as for firms in other service sectors (Tether, 2005). Research also suggests that it is seldom possible to protect knowledge used in innovation processes via patents or other types of intellectual property rights (Mendonca *et al.*, 2004). It has also been found that knowledge used in services, including tourism, are generally more difficult to modularize compared to knowledge used in manufacturing (Aas & Pedersen, 2013).

These characteristics of knowledge and innovation practices in tourism may complicate outbound open innovation (Aas & Pedersen, 2016). Tacit knowledge may for example in general be difficult to transfer to other actors without a prior resource demanding conversion to explicit knowledge (Nonaka, 1994). The pecuniary type of outbound open innovation may be particularly difficult to implement due to the fact that knowledge used in services is difficult to modularize, protect and put value on (Mendonca *et al.*, 2004; Aas & Pedersen, 2013). Thus, it has been hypothesized that "successful implementation of open service innovation is more likely for non-pecuniary rather than pecuniary open innovation" (Aas & Pedersen, 2016, p. 312).

However, with some exceptions (e.g., Aas, 2016), there are few empirical studies that shed light on the phenomenon of outbound open innovation in tourism and for this reason this explorative study was undertaken.

3. Method

It has been argued that a case study design may be preferable when we want to explore and achieve in-depth knowledge about a phenomenon that we have limited prior knowledge of Yin (2014). Therefore, to explore how outbound open innovation processes in tourism look like and to address the explorative research questions raised in this chapter a case study approach was chosen.

To get in-depth knowledge we chose to focus on and observe an innovation project called INSITE during its implementation. The aim of INSITE was to improve the operational decision-making processes and subsequently the financial performance of the participating tourism firms by implementing and exploiting new information technologies, such as beacons and data visualization panels. Thus, INSITE may be categorized as a process innovation project.

One firm, here called Firm A, that operated an amusement park providing a variety of attractions including a zoo, numerous theatre shows, a water park, themed accommodation as well as other experiential services, took the initiative to establish the INSITE project in 2016. Although this amusement park was open 365 days a year, it had its main earnings from mid-June to mid-August. In 2016 the firm had about 70 permanent employees and 50 part-time employees, but in the period mid-June to mid-August the staff was supplemented by about 1,000 seasonal workers (Sveindal & Amtrup, 2016). The park had more than 1,000,000 visitors in 2016. The huge size of the park, especially during peak season, complicated daily operations of the park and perhaps especially the process of identifying areas for continuous improvement. As a remedy against this situation the INSITE project was initiated.

The management of Firm A believed that the INSITE project would be relevant also for other firms in the tourism industry and invited other firms, mainly form the same region, into the project. Ten tourism firms accepted the invitation. The participants came from many different tourism sub sectors, and included transportation firms, hotels, a restaurant, other amusement parks, an event agency, a museum and a ski resort. In addition a consultancy firm, a

Table 1. Tourism firms participating in the INSITE project.

Firm	Sector	Annual turnover (the year 2016)
A	Amusement park	429 MNOK
B	Amusement park	13.2 MNOK
C	Hotel	48.3 MNOK
D	Event agency	11.3 MNOK
E	Ski resort	30.4 MNOK
F	Transportation	703 MNOK
G	Museum	12.6 MNOK
H	Restaurant	22.6 MNOK
I	Transportation	133 MNOK
J	Amusement park	25.1 MNOK
K	Hotel	3.4 MNOK

technology provider, a destination marketing organization as well as researchers joined the project. The project was in part funded by the participating tourism firms and in part by the Research Council of Norway. Table 1 provides an overview of the tourism firms participating in the INSITE project and their role.

The INSITE project was chosen as a case for this study for several reasons: First, the project was aiming to considerably change the processes of the participating firms and could therefore be characterized as an innovation project. The fact that the project had received support from the user-driven innovation program in the Research Council of Norway also support that it would be fair to say that the INSITE project could be called an innovation project. Second, more than one tourism firm participated in the project, increasing the likelihood of identifying flows of knowledge between the participating tourism firms during the innovation process. Thus, an in-depth study of the knowledge flows in the INSITE project constituted a particularly good opportunity to learn about what type of knowledge that is transferred in outbound open innovation processes in tourism, how the knowledge is transferred and what the advantages and disadvantages are.

As researchers we were allowed to observe the activities in the INSITE project during its implementation. We did this by

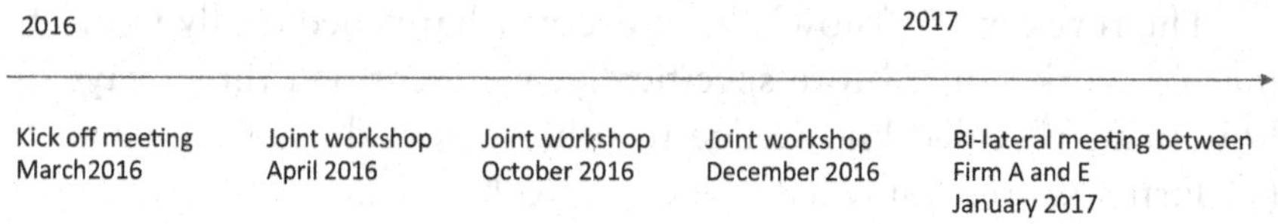

Figure 1. Activities observed during the implementation of the INSITE project.

participating as observers on all joint workshops in the project. In addition we participated as observers in a bi-lateral meeting between Firms A and E. The activities observed were taped with a tape recorder and transcribed. Figure 1 illustrates the activities observed during the study pinpointed on a timeline.

In addition to the observations we conducted an in-depth interview with the CEO of Firm A. He was the initiator of the project as well as the project owner, and very central during the implementation of the project. An interview guide was used during this interview and the interview was taped and transcribed.

To address the RQs the transcribed data were analyzed in the following manner: We thoroughly read through the material and identified the instances where one tourism firm shared innovation relevant knowledge with one or several other partners. We noted what type of knowledge that was transferred and how the knowledge was transferred, and searched for patterns in the data. We also searched through the material for why data were transferred. The main results of this analysis exercise are now reported.

4. Finding

4.1. *How outflows of knowledge were transferred*

During the period that we observed the activities in the INSITE project, we did not identify any knowledge flows of the pecuniary type. The tourism firms that shared knowledge with other actors during the project did not aim to commercialize their knowledge in any way. Thus, the knowledge was not sold or licensed out, but instead revealed for free without a financial reward.

The revealing of knowledge moreover happened orally in meetings or workshops. More specifically, we identified three ways of how firms revealed knowledge to other actors during the project: (1) Firms shared knowledge with all other firms participating in the project during joint workshops, (2) firms shared knowledge with other firms participating in the project in bi-lateral meetings, (3) firms shared knowledge with the consultancy firm, the technology provider, the destination marketing organization or the researchers during the project, and these actors shared the knowledge with the other actors in bi-lateral meetings or in joint workshops.

4.2. *The types of knowledge transferred*

4.2.1. *Knowledge transferred in joint workshops*

During the joint workshops arranged in the project all participating tourism firms revealed knowledge of some kind to the other actors. The knowledge shared in these workshops was often related to sharing practical challenges experienced during the implementation of the project and how these were solved. Examples include that Firm F presented challenges with using beacons in areas outside mobile coverage, and how they had solved this, Firm K explained how they had experimented with alternative technologies to save costs and Firm A explained how many push messages they could send to guests before they logged out from the app. The knowledge shared in these joint workshops often had a relative general character. The firms did not share details on how things had been implemented, but rather a general overview that could be useful for the other firms especially in the initial stages of the project.

In addition to sharing practical experiences the initiator of the project (Firm A) also shared the firm's more strategic ambitions with the INSITE project, including how the technology to be implemented could contribute in achieving the firm's strategic goals. For example during one workshop the CEO of Firm A stated:

> *"Insight about our customers is needed to optimize our production. We simply need to know what attractions they visit, what matters to them, how satisfied they are with the attractions. And that is where the app and the beacons comes in. We can see them with the bluetooth, or we can talk to them through the app (...). And then we get an overview of how we have produced, right. (...) Then I can use this information to make the right investments, and make the right production. Perhaps I can take away some things from our production without reducing the customer satisfaction. For example now we have 18 animal presentations per day during the peak season. If 15 were enough, we would need so much resources. But without the beacons system I have no idea if it is smart to reduce to 15 (...)."*

4.2.2. *Knowledge transferred in bi-lateral meetings*

As mentioned the knowledge shared in the joint workshops was relatively general. More detailed knowledge was shared in bi-lateral meetings between participants. We observed a bi-lateral meeting between Firms A and E (a ski resort), and during this meeting Firm A revealed the following types of knowledge: (1) detailed information about technical solutions and vendors, (2) detailed information about need for improvements, and (3) detailed information about the resources needed to develop and implement the system. Table 2 illustrates these types of knowledge with relevant quotes.

In addition to sharing knowledge resulting from its own experiences Firm A also engaged in a discussion with Firm E on how it should solve the specific challenges facing this particular firm. During this discussion Firm A tried to understand Firm E's situation and propose solutions. The CEO of Firm A for example stated:

> *"What is it that makes your customers want to use the app? What useful information do you as a firm have that the skier would find useful? (...) For example perhaps a skier would like to know where*

Table 2. Knowledge transferred in bi-lateral meetings.

Type of knowledge	Illustrative quotes
Detailed information about technical solutions and vendors	"We have chosen to deploy 162 beacons in the park." "So that means that if we want to measure sales in the shop, the beacon must be placed right above the cash register. It must also have a radius of 50 centimeters, otherwise you will not know if they have bought or not (...). But if you're going to measures guests visiting the lion tribune you may have a radius of 30 meters, because it captures the entire tribune, right (...)." "Firm NN [anonymized] is definitely a firm that can have the skills to help you with that (...), and nn [anonymized] is a great and sympathetic guy (...), so I'm not afraid to recommend him (...)."
Detailed information about need for improvements	"So we're going to further develop the app by integrating sale of tickets in the app (...). That means that in the future you as a customer will more or less be forced to use this app (...)."
Detailed information about the resources needed to develop and implement the system	"The app we have now costed some hundred thousand. But we're going to make a much more comprehensive app, we're going to invest more in the app. There is no doubt that we should have added more features in the app, and definitely tickets on passage control."

the queue is (...). I guess this information is available and can be pushed? (...) The kind of service messages that really matter to the guest, will determine if they will download the app. (...)."

4.2.3. *Knowledge transferred via a third party (e.g., consultants and researchers)*

Both the consultants and the researchers in the project consortium participated in many meetings with firms during the implementation of the project, both as advisers and observers. During these meetings a lot of knowledge was transferred from the firms to the researchers and consultants. Some of this knowledge was sensitive from a commercial point of view and should not be shared with other firms, but some knowledge was not sensitive and this knowledge was often disseminated to other firms by the researchers and consultants either in joint workshops or in bi-lateral meetings between the consultants/researchers and individual firms. During one joint workshop one of the consultants for example stated:

> *"I have experienced that when placing beacons, start with a map of the area and place the different beacons in different areas, name the different areas, and think about what is the purpose of each beacon and what range they need (...)."*

4.3. *Why knowledge was transferred*

We identified quite a few outflows of knowledge from all participating firms during the INSITE project. However, perhaps unsurprisingly on this point the initiator of the project, Firm A, excelled. We therefore interviewed the CEO of Firm A to better understand why the Firm decided to transfer knowledge to other actors during the implementation of the INSITE project. During this interview several reasons were identified. First, the informant argued that sharing knowledge with other parties will increase the likelihood of learning from other firms and in the next turn improve your own firm. The informant stated:

> *"(...) It is really not risky to share knowledge (...). You're not getting worse because someone else gets better, it really just sharpen you up. (...) So it has become quite natural for us to tell others what we are doing. (...)."*

Second, the informant suggested that sharing knowledge with others affect the image and reputation of the firm in a positive manner. The informant said:

> *"I have become much more aware of the fact that we are building our reputation when we share knowledge. People see that we are*

willing to share and that is something that seems positive, they become positively tuned. They will usually not say anything negative about us due to the fact that they are positively tuned."

Third, the informant argued that the sharing of knowledge increases the total market size. He said:

"Our sharing philosophy has inspired other competing firms to establish 'the children's zoo' and 'the little animal park' (...), but I cannot see that we will lose anything at all, as a consequence of this. These initiatives will only attract much more people to come to this region. The market will increase. The cake will become bigger. The attractiveness of the region will increase. And if we are unable to develop ourselves quickly enough, we will lose regardless of our sharing policy (...)."

The informant had a very positive attitude towards sharing of knowledge during the in-depth interview. However, during the observation of the bi-lateral meeting between Firm A and Firm E we also experienced that there are some situations where sharing of knowledge may not be recommended. When discussing whether Firm E should share their knowledge related to beacons technology with the local destination marketing office, the CEO of Firm A stated:

"Be careful when you share knowledge with so-called neutral partners. (...) If they become skilled they can suddenly have the customer insights that you really should have. It's okay to do this project in partnership, but the project consortium should consist of other businesses who have their own 'cash registers', and not neutral advisors (...). It just does not work (...). The danger is if the neutral partners start communicating with the guests. You should have all the communication with the guests. You can also have business partners in that communication. That is interesting because you are going to have a return with someone, right? They are the ones you should communicate together with, because they

will be able to fix problems with their offer. Those who are neutral can never fix anything, they can only say that nothing was done."

5. Discussion and Conclusions

The experiences from the INSITE project confirm that outbound open innovation is a relevant and feasible type of open innovation for tourism firms. By observing the implementation of the project we identified numerous outflows of knowledge from tourism firms to other tourism firms in the project consortium. Many different types of knowledge were transferred during the outbound open innovation processes in the case under study. The firms transferred knowledge resulting from their own experiences and in addition they engaged in discussions on how other firms could solve their challenges. All outflows of knowledge had a non-pecuniary nature. This does not necessarily mean that pecuniary open innovation is irrelevant in tourism, but it supports the idea suggested by prior research (e.g., Aas, 2016) that the non-pecuniary type of outbound open innovation is more common than the pecuniary type in this sector, in part due to the tacitness of knowledge used in innovation processes in this sector (Cooper, 2006).

The findings indicate that tourism firms decide to reveal knowledge for free to other firms for about the same reasons as firms in other sectors decide to reveal knowledge (Allen, 1983; Dahlander & Gann, 2010; Wang & Zhou, 2010): They believe that the revealing of knowledge will improve their image, they believe that the revealing of knowledge will increase the likelihood of learning from other firms, and they believe that the revealing of knowledge can have positive consequences for the sector as a whole. The latter reason seems to be particularly relevant in tourism since tourism firms from different sub-sectors are conjoined in natural value networks and are mutually dependent on each other's achievements.

The findings also shed light on how non-pecuniary outflows of knowledge may be transferred in tourism. The findings suggested that the knowledge may be revealed in joint workshops where many firms participate or in bi-lateral face-to-face meetings between firms.

It is notable that the knowledge was transferred in arenas where the revealers of knowledge and the recipients of knowledge had a chance to discuss. In some cases we also identified that the knowledge was transferred via researchers or consultants. The relatively high degree of tacitness of much of the knowledge to be shared or transferred may explain both why workshops or face-to-face meetings seemed to be well functioning arenas for non-pecuniary knowledge revealing, and why the use of "knowledge mediators" was useful in some cases. According to Nonaka (1994)'s theory in the area of knowledge management, tacit knowledge must be transformed to explicit knowledge before it is possible to transfer it to other actors. Although these theories have been debated and criticized (e.g., Gourlay, 2006), the lessons from the INSITE project suggest that knowledge becomes more explicit and thus transferable when the recipients of knowledge have the opportunity to discuss with the revealers of knowledge, and that knowledge mediators such as consultants and researchers may play a valuable role in helping firms make the knowledge explicit and transferable.

We observed that most of the knowledge flows identified happened within the project consortium. This, as well as the results from the in-depth interview, substantiates that a certain level of trust between the firms revealing knowledge and the firms receiving knowledge is needed in outbound open innovation processes in tourism. In particular this applies for the situation where knowledge is transferred via a knowledge mediator (e.g., consultant or researcher). The firm revealing knowledge must be sure that the third party does not use the knowledge in a way that can influence the relationship between the revealing firm and its customers. The importance of trust in open innovation processes has to some degree been discussed and acknowledged in the literature (e.g., Chatenier *et al.*, 2010; Lee *et al.*, 2010), and the findings of this case study indicate that trust between all actors involved is a particularly important antecedent of outbound open innovation processes in tourism. Further research should investigate what the sources of trust between actors are and how firms build sufficient trust to other actors.

Further research is also needed to confirm and elaborate on the findings of this study. The chapter is based on the in-depth study of one case and it may be argued that this research design has limitations related to generalizability. We therefore suggest that continued qualitative exploration of innovation activities in tourism is needed to enhance our understanding of how outbound open innovation processes are managed in this sector.

Acknowledgments

This work was supported by the Research Council of Norway.

References

Aas, T. H. (2016). Open service innovation: The case of tourism firms in Scandinavia. *Journal of Entrepreneurship, Management and Innovation*, 12(2), 53–76.

Aas, T. H., & Pedersen, P. E. (2013). The usefulness of componentization for specialized public service providers. *Managing Service Quality*, 23(6), 513–532.

Aas, T. H., & Pedersen, P. E. (2016). The feasibility of open service innovation. In: A.-L. Mention and M. Torkkeli (eds.), *Open Innovation: A Multifaceted Perspective.*World Scientific Publishing, London.

Allen, R. C. (1983). Collective invention. *Journal of Economic Behaviour and Organization*, 4(1), 1–24.

Chatenier, E. D., Verstegen, J. A., Biemans, H. J., Mulder, M., & Omta, O. S. (2010). Identification of competencies for professionals in open innovation teams. *R&D Management*, 40(3), 271–280.

Chesbrough, H. (2003). *Open Innovation: The New Imperative for Creating and Profiting from Technology.* Harvard Business School Press, Boston, MA.

Chesbrough, H. (2010). *Open Services Innovation: Rethinking Your Business to Grow and Compete in a New Era.* Jossey-Bass, San Francisco.

Chesbrough, H., & Bogers, M. (2014). Explicating open innovation: Clarifying an emerging paradigm for understanding innovation. In H. Chesbrough, W. Vanhaverbeke, and J. West (eds.), *New Frontiers in Open Innovation.* Oxford University Press, Oxford, pp. 3–28.

Chesbrough, H., & Rosenbloom, R. S. (2002). The role of the business model in capturing value from innovation: Evidence from Xerox Corporation's technology spin-off companies. *Industrial and Corporate Change*, 11(3), 529–555.

Chesbrough, H., Vanhaverbeke, W., & West, J. (2006). *Open Innovation: Researching a New Paradigm*. Oxford University Press, London, England.

Cooper, C. (2006). Knowledge management and tourism. *Annals of Tourism Research*, 33(1), 47–64.

Dahlander, L., & Gann, D. M. (2010). How open is innovation? *Research Policy*, 39, 699–709.

Droege, H., Hildebrand, D., & Forcada, M. A. (2009). Innovation in services: Present findings, and future pathways. *Journal of Service Management*, 20(2), 131–155.

Fosfuri, A. (2006). The licensing dilemma: Understanding the determinants of the rate of technology licensing. *Strategic Management Journal*, 27(12), 1141–1158.

Gambardella, A., Giuri, P., & Luzzi, A. (2007). The market for patents in Europe. *Research Policy*, 36(8), 1163–1183.

Gourlay, S. (2006). Conceptualizing knowledge creation: A critique of Nonaka's theory. *Journal of Management Studies*, 43(7), 1415–1436.

Helfat, C. E. C. (2006). Book review of open innovation: The new imperative for creating and profiting from technology. *Academy of Management Perspectives*, 20(2), 86.

Henkel, J. (2006). Selective revealing in open innovation processes: The case of embedded Linux. *Research Policy*, 35(7), 953–969.

Hjalager, A. M. (2010). A review of innovation research in tourism. *Tourism Management*, 31(1), 1–12.

Hu, Y., McNamara, P., & McLoughlin, D. (2015). Outbound open innovation in bio-pharmaceutical out-licensing. *Technovation*, 35, 46–58.

Huizingh, E. K. (2011). Open innovation: State of the art and future perspectives. *Technovation*, 31(1), 2–9.

Lee, S., Park, G., Yoon, B., Park, J. (2010). Open innovation in SMEs — An intermediated network model. *Research Policy*, 39, 290–300.

Mendonca, S., Santos Pereira, T., & Godinho, M. M. (2004). Trademarks as an indicator of innovation and industrial change. *Research Policy*, 33, 1385–1404.

Nerkar, A. (2007). *The folly of Rewarding Patents and Expecting FDA Approved Drugs: Experience and Innovation in the Pharmaceutical Industry*. Working Paper, Kenan-Flagler Business School, University of North Carolina at Chapel Hill.

Nonaka, I. (1994). A dynamic theory of organizational knowledge creation. *Organization Science*, 5(1), 14–37.

Srholec, M., & Verspagen, B. (2012). The Voyage of the Beagle into innovation: Explorations on heterogeneity, selection, and sectors. *Industrial and Corporate Change*, 21(5), 1221–1253.

Sveindal, H. M., & Amtrup, J. (2016). Den levende parken: Dyreparken gjennom 50 år. Oslo: Ena. (The living park: The zoo through 50 years.)

Tether, B. S. (2005). Do services innovate (differently)? Insights from the European innobarometer survey. *Industry and Innovation*, 12(2), 153–184.

Wang, P., & Zhou, Y. (2010). The role of outbound-revealing open innovation: Theoretical extension and case study. In: *2010 IEEE International Conference on Industrial Engineering and Engineering Management* (*IEEM*), IEEE, pp. 1890–1892.

West, J., & Bogers, M. (2014). Leveraging external sources of innovation: A review of research on open innovation. *Journal of Product Innovation Management*, 31, 814–831.

West, J., & Bogers, M. (2017). Open innovation: Current status and research opportunities. *Innovation: Organization and Management*, 19(1), 1–8.

Yin, R. K. (2014). *Case Study Research: Design and Methods* (5th edn.). SAGE, Los Angeles, California.

https://doi.org/10.1142/9789811234491_0009

Chapter 9

Opening Up the Service Innovation Process Towards Ordinary Employees in Large Service Firms

Tor Helge Aas

School of Business and Law, University of Agder, Campus Kristiansand, Universitetsveien 25, 4630 Kristiansand, Norway

NORCE Norwegian Research Centre, Nygårdsgaten 112, 5008 Bergen, Norway

tor.h.aas@uia.no

Abstract. Most research on how the intellectual resources needed to succeed with innovation are managed has focused on the management of external intellectual resources in open innovation processes, whereas the management of the intellectual resources that ordinary employees possess has not received the same attention. This chapter addresses this literature gap by exploring employee involvement in service innovation processes qualitatively. This is done by following and observing the implementation of a project aiming to increase a large service firm's innovativeness through higher and more focused employee involvement. Our findings suggest that successful implementation of employee involvement in this context is associated with the support of managers at all levels, the establishment of arenas where employees can collaborate, the establishment of a transparent system for screening and selection of ideas, and the existence of professional innovation management employees that are able to facilitate and lead the development of ideas.

Keywords. Service innovation; employee-driven innovation; high-involvement innovation; open innovation.

1. Introduction

To succeed with innovation activities firms need access to intellectual resources, including educational, cultural, and experiential knowledge and skills (Froehle & Roth, 2007). In recent years, innovation management research has focused predominantly on how intellectual resources outside the boarder of the individual firm may be used to accelerate internal innovation in so-called open innovation processes (Huizingh, 2010). Suppliers, business partners and customers represent potential external intellectual resources that may be valuable in open innovation processes (Ulrich & Ellison, 2005), and the literature discusses both advantages and disadvantages with the use of external resources (Dahlander and Gann, 2010), as well as the associated management challenges (van de Vrande *et al.*, 2009; West & Bogers, 2014).

While the involvement of external intellectual resources has received much attention in innovation management research recent years, the involvement of the intellectual resources that ordinary internal employees possess has not received the same attention (Kesting & Ulhøi, 2010). However, based on the assumption that employees often have hidden abilities for innovation (Ford, 2001) it may be beneficial for firms to open up the innovation processes internally towards ordinary employees. This may particularly be the case in large firms where knowledgeable employees often work in organizational or functional silos that do not necessarily collaborate much (Gulati, 2007).

It has been argued that the involvement of employees in innovation processes may be particularly valuable in service firms (Engen & Magnusson, 2015; de Jong *et al.*, 2003). Ordinary employees in service firms often have direct interactions and contact with customers (Skålén *et al.*, 2015), and they therefore obtain a unique knowledge both about the customers' needs and the firm's strengths and weaknesses that can be applied to generate innovative new service ideas (Lages & Piercy, 2012).

Many different types of processes can lead to the development of new services: Service innovations can be the result of a formal process that is initiated, implemented and controlled by the management of the firm (de Jong & Vermeulen, 2003), but it can also be the result of an informal bricolage process where service employees gradually change their work practices (Salunke *et al.*, 2013). In bricolage processes ordinary employees obviously play the main role. However, even if these processes often happen without the direct involvement of managers, managers may play an important role in creating an organizational culture that appreciate or encourage bricolage (Fuglsang & Sørensen, 2011).

This chapter, however, focuses on the more formal management-driven service innovation processes in firms. How ordinary employees should be involved in these processes is more uncertain. Some research has focused on the effects of involving ordinary employees in service firms in management-driven service innovation activities (e.g., Melton & Hartline, 2010; Ordanini & Parasuraman, 2011), but the results of this stream of research is inconclusive. It has therefore been argued that more qualitative research is needed on how ordinary employees in service firms are involved in management-driven service innovation activities (Engen & Magnusson, 2018), and what the managerial antecedents are (Kesting & Ulhøi, 2010). As a response to this call for more research we raise the following research question (RQ) in this chapter: What are the managerial antecedents of successful employee involvement in management-driven service innovation processes in large service firms?

An in-depth qualitative case study where we investigated how managers of a large service firm involved ordinary employees in their management-driven innovation processes was conducted to address this RQ. The chapter is organized in the following way: In the next section we review the relevant literature on the antecedents of employee involvement in innovation processes in large service firms.

The research method is discussed in Section 3. Section 4 reports the findings and in the last section, Section 5, the findings are discussed and conclusions are provided.

2. Literature Background

In part due to technological progress workplaces are becoming more and more complex, and in most firms human capital is now an important success factor not only at top management level, but also for "ordinary" employees (Kesting & Ulhøi, 2010). Consequently, employee participation in firms' management processes is becoming an increasingly important organizational asset (e.g., Herstein & Mitki, 2008; Addison & Belfield, 2004). Employee participation in general has also been, and still is, a hot topic in management research, and a recent literature review (Bakker & Demerouti, 2008) concludes that employees that are involved in their firm's management processes are more productive, and more willing to go the extra mile for their firm.

Although there has been relatively much research on employee involvement in general management processes, less research focuses specifically on the involvement of employees in firms' innovation processes. Employees' innovative behavior can be defined as "an employee's intentional introduction or application of new ideas, products, processes, and procedures to his or her work role, work unit, or organization" (Yuan & Woodman, 2010, p. 324). Innovation processes with high employee involvement is sometimes referred to as employee-driven innovation (EDI) processes, where EDI means "the generation and implementation of significant new ideas, products and processes originating from a single employee or the joint efforts of two or more employees who are not assigned to this task" (Kesting & Ulhøi, 2010, p. 66). Another related term that is sometimes used to describe innovation processes with high employee involvement is "high-involvement innovation" (HII) (Bessant & Caffyn, 1997). HII may be defined as "an approach whereas many employees as possible are sought to be involved in participating at any degree in the innovation process" (Hallgren, 2009, p. 50).

The involvement of ordinary employees has been suggested to be essential for service innovation (e.g., Grönroos, 2008; Melancon *et al.*, 2010). Service innovation refers to the development and implementation of new service solutions or experiences (den Hertog *et al.*, 2010). Such innovations may include changes in the service concept, client interaction channel, service delivery channel and/or the supporting technology (den Hertog, 2000). As noted in the introduction of this chapter many different types of processes can lead to the development of new services. New services can be the result of formal visible management-driven processes (de Jong & Vermeulen, 2003), or informal invisible bricolage processes (Fuglsang & Sørensen, 2011). In practice the distinction is not always completely clear since many service innovations are the result of a process that is a combination of bricolage processes and formal processes and driven both by managers and employees (Aas *et al.*, 2016). Nevertheless, conceptually it is possible to draw a line between the emerging bricolage processes mainly driven by employees and the more formal innovation processes initiated and mainly driven by management.

In this chapter, we have chosen to focus on visible formal service innovation processes that are initiated, led, controlled and mainly driven by the management of the firm since the role that ordinary employees play in these processes is particularly unclear. Management-driven service innovation processes typically start with a search stage where the firm searches for new service ideas and they end with an implementation stage where the new service is launched in the market (de Jong *et al.*, 2003). How the process is managed varies and depend upon the type of service being developed (e.g., Jaakkola *et al.*, 2017). The process can consist of a set of predefined stages and gates, in the same manner as what is prescribed to product innovation processes (Cooper, 2008), but in many cases the process is more iterative and ad-hoc than a typical product innovation process (e.g., Hipp & Grupp, 2005; Aas, 2017). The conceptual complexity of service innovation often also imply that organizational processes and structures are changed during the process (de Jong *et al.*, 2003). This may require that managers from many different functional areas, including operations, are involved in the service innovation process (e.g., Aas *et al.*, 2017), but whether this also includes the involvement of ordinary employees from different functional areas is not clear.

Existing research on the involvement of employees in innovation processes typically focus on the behavior of employees (e.g., de Jong & den Hartog, 2010; Scott & Bruce, 1994) or their creativity (e.g., Bharadwaj & Menon, 2000; Gilson & Madjar, 2011), and has mostly focused on the importance of implementing a focused strategy, a specific organizational structure and a supportive culture to succeed with the involvement of employees (e.g., de Jong & Kemp, 2003).

Byrne *et al.* (2009), however, call for more research on the relationship between managerial behavior and involvement of ordinary employees in innovation processes. This call is echoed by Kesting and Ulhøi (2010) who state that research on involvement of ordinary employees in innovation processes is still in its infancy, and that more knowledge especially about the managerial antecedents is much needed. This gap also applies for the specific case of service innovation (Engen & Magnusson, 2018) and this chapter therefore aims to address this gap.

3. Method

To explore how employees are involved in service innovation processes, and how the involvement is managed, in large service firms, and answer the research question, we deployed a qualitative case study approach, since this approach arguably has advantages when the phenomenon to be studied is not well understood and where the variables are still unknown (Johnson & Harris, 2003). We purposely selected a large Scandinavian provider of IT services as a case organization. The case organization had around 10,000 employees and was selected primarily because it was running an internal project aiming to increase its innovativeness through higher and more focused employee involvement.

This internal project in the case organization, hereinafter called the HII project, lasted for approximately 8 months and was implemented in three steps: A new software portal for capturing innovative ideas from employees was implemented in step 1, a new method to screen and select a shortlist of the most valuable ideas with the assistance of employees was implemented in step 2, and a new method for detailing and specifying the short-listed ideas together with a group of employees collaborating in a workshop was implemented in step 3.

As researchers we were allowed to follow and observe the entire implementation of the HII project. Thus, the research design may be characterized as a natural field experiment, where the antecedents of successful employee involvement were explored. We used several methods of data collection during the process including in-depth interviews and observation: We started the data collection by observing the implementation of a one-day workshop that was arranged relatively late in the HII project (step 3). In this workshop, the originators of 12 innovation ideas (ordinary employees), as well as 10 managers participated. During the workshop the idea originators and managers worked together to make the 12 ideas more concrete, and they also ranked the ideas. During the workshop we observed and took notes on observations related to the research question, and insights from the workshop were used to select informants for further in-depth interviews.

We continued the data collection by interviewing the senior manager that took the initiative to start the HII project in the case organization. He suggested that we should also interview three other managers that had been heavily involved in the implementation of the HII project. The interviews of the four managers were conducted relatively late in the HII project, after step 3, to ensure that the managers had gained experience prior to the interviews. The interviews were conducted in a semi-structured manner. An interview guide was used during the interviews (see Appendix A). The interviews were organized according to the process of the HII project and the informants were asked open questions about their experiences and lessons learned during the HII project, and their answers were followed up by more detailed questions about their role.

After interviewing the managers we interviewed three ordinary employees that had submitted ideas during the HII project (step 1). These employees had also participated in the workshop that we observed, and were selected as informants due to the fact that they were particularly active in the workshop, indicating that they had the innovative behavior that the managers wanted to stimulate. We expected that much could be learned about managerial antecedents by approaching these persons. An interview guide was also used during these interviews (see Appendix B). All interviews lasted approximately one hour and were transcribed. This resulted in a very rich data set, well suited to suggest answers to the research question.

To make sense of the empirical data, we followed an inductive approach: The data was examined in light of the research question by searching for what antecedents of successful employee involvement in innovation processes that appeared in the interview transcripts both from the interviews with managers and from the interviews with ordinary employees. The antecedents appearing from this exercise were then grouped into four categories. The empirical findings are now reported.

4. Findings

During stage 1 of the HII project the case organization implemented a new software tool for capturing innovative ideas from employees. The HII project management informed all employees in the case organization about this new tool on the firm's intranet. The employees were given a reasonable deadline (about two months) to submit ideas related to a specific strategic challenge defined by the management of the firm. When the deadline expired approximately 100 ideas had been submitted. Several of our informants, both managers and ordinary employees, stated that this number was surprisingly low.

In the in-depth interviews with managers and employees we were given several explanations why the 'campaign' did not engage employees as much as expected. A line manager for example highlighted that the campaign could have been marketed better:

> *"In a way I think employees in our firm understand that innovation is important for the firm, but generally speaking it is difficult to get people to really engage, to spend time and resources on defining new ideas. I think perhaps people do not think there are enough incentives to engage. The implementation of the new software tool did not change this fundamental problem. It is in a way not enough, I think, to send an email to the employees and tell them that a new portal has been implemented and that they should submit new ideas into this portal ... More effort is needed to create engagement among employees ..."*

Several informants also highlighted that it was not clear for them that the initiative was supported by the top management and rooted in the firm's strategy, and that the top management really wanted employees to focus on developing new and innovative ideas. One line manager for example stated:

> *"To my experience there was some information about this campaign on the intranet etc ... but what I think is very important to succeed with innovation is that the initiative is anchored in the top management, and that it is evident for all the line managers and employees that this initiative in a way is on the top of the mind of the top management. In my view this is not evident in our firm. Perhaps our CEO think innovation is important, but I am not sure. It seems as if this has been delegated to some business developers in the line organization. This is perhaps correct, but I think the CEO should have been clearer in explaining that this is important..."*

Several informants highlighted that they would have liked the organization to establish arenas where employees were allowed to spend time and collaborate on the development of ideas. One employee for example stated:

> *"I know that in other firms, for example Google, employees are allowed to spend time on idea development... We do not have this*

system here... That is a problem from an innovation point of view I think. People are very busy with operations during their work days, and the time available to think about other things is very limited."

This lack of both time and arenas to discuss new ideas was confirmed by several managers. One manager for example stated:

"I did not explicitly say that the employees in my department were not allowed to work on new ideas during their work hours, but I made it clear that the daily operations should have priority. It is important that we deliver what we have promised to our customers."

Thus, the explanations on why the "campaign" did not engage employees as much as expected may be summarized in three points: (1) it was not communicated clearly enough that the initiative was supported by the top management and rooted in the firm's strategy, (2) incentives related to innovation were lacking, and (3) the firm did not establish arenas for collaboration among employees.

During stage 2 of the EDI project all employees in the case organization were given the opportunity to vote for the ideas submitted. This voting process also suffered from a lack of employee engagement, and the management only partially took the voting results into account when they decided what ideas to shortlist. However, our interview results suggested that employees understood and accepted that the voting results were only one of several factors when deciding what ideas to shortlist. One informant (employee) for example stated:

"It is very understandable that the voting results were one of many factors when evaluating what ideas to shortlist, especially when so few employees actually voted..."

As a result of stage 2, 12 ideas were shortlisted. In stage 3 of the HII project a workshop was arranged, where the originators of these 12 shortlisted ideas, as well as about 10 line-managers participated. During the workshop the idea originators and managers worked together to make the ideas more concrete. Our findings suggested that both ordinary employees and managers found the workshop very useful. Especially the collaboration between employees with different backgrounds resulted in significant improvement of the ideas. However, several informants stated that they would have appreciated if a similar event could have been arranged earlier in the process. One employee for example stated:

> *"The workshop was a very good experience. During the workshop we were able to improve the ideas very much ... It was very valuable to discuss with employees from other departments ... I only wish this workshop had been arranged earlier in the process."*

Some idea originators also expressed that they would have appreciated if dedicated innovation experts could have been involved earlier in the process both to assist in the concretization and specification of the ideas, and also to manage the idea development process. The following statement from one idea originator exemplifies this:

> *"It was to some degree a coincidence that I became the originator of this idea. In fact this was an idea that originally came from a colleague that had left the company, so I just submitted it to the system. I think it is important that employees should be allowed to submit ideas even if they do not have detailed knowledge of how to develop the idea into a new product or service ... In this process I think the company put too much responsibility on the idea originator. I would have liked if some professional people could take over the idea and manage the development process ..."*

5. Discussion and Conclusions

By using rich empirical case study data from a large IT service provider the chapter addresses the literature gap related to the managerial antecedents of successful employee involvement in service innovation processes in large firms. Although the study has an explorative nature, the findings contribute considerably to our knowledge by providing a basis for discussing both managerial prescriptions and the development of a theory of HII managerial antecedents.

Based on our empirical findings we suggest that the managerial antecedents are categorized in four broad categories: (1) Incentive systems, (2) Collaboration arenas, (3) Screening and selection system, and (4) Professional innovation employees. Each group of antecedents is now discussed:

Our first category, incentive systems, relates to our finding that ordinary employees typically need to be absolutely sure that the senior management wants them to spend time on innovation activities before they engage and participate in innovation activities. In other words, the employees need to be sure that the effort will be supported and appreciated by the management. This finding is fully in line with the results of research on innovative behavior. The seminal study of Scott and Bruce (1994), for example, found that support for innovation was an important determinant for innovative behavior, and this has been confirmed by a more recent meta-study (e.g., Hammond *et al.*, 2011). In addition to this our findings also indicate that the implementation of formal and transparent incentive systems may be an efficient way for the management to demonstrate that they really support and appreciate employees' efforts and involvement in innovation activities. This has also been suggested by the conceptual literature on employee-driven innovation (e.g., Kesting & Ulhøi, 2010). Based on our findings it is, however, not possible to conclude how exactly the incentive systems should be designed in different contexts. We suggest that future research should focus more specifically on that. Nevertheless, our findings do enable us to offer Proposition 1:

Proposition 1. *The support of managers at all levels and the implementation of formal incentive systems are associated with the successful implementation of high-involvement service innovation in large firms.*

Our second category of managerial antecedents is called collaboration arenas. Previous research has indicated that managers that provide relevant resources to employees stimulate their innovative behaviour (e.g., Scott & Bruce, 1994). Our findings support this, and suggest that the establishment of arenas where employees can collaborate is particularly relevant to succeed with HII in large service firms. The nature of service innovation may explain this finding. Service innovation has a high contextual complexity and often involve changes in both the service concept and the organizational processes and structures (de Jong & Vermeulen, 2003). Thus, collaboration between employees from different functional areas may be particularly important in service innovation, especially in large and complex organizations. Hence, we suggest Proposition 2:

Proposition 2. *The establishment of arenas where employees can collaborate on the development of ideas is associated with the successful implementation of high-involvement service innovation in large firms.*

Our third category of managerial antecedents is called screening and selection system. We found that the establishment of a transparent system for screening and selection of innovation ideas was important to stimulate the involvement of employees in innovation processes. This category of managerial antecedents is not discussed explicitly in previous literature on HII (e.g., Kesting & Ulhøi, 2010). It has, however, been suggested in the literature that managers' ability to build a relationship based on trust and respect with their employees is positively correlated with the innovative performance of employees (e.g., Hammond *et al.*, 2011). The implementation of transparent systems for screening and selection of innovation ideas may be one practical way for managers to build a trustful relationship with employees, due to the fact that such systems will enable employees to see that ideas are taken seriously and treated in a respectful and fair manner by the management of the firm. Hence, we offer Proposition 3:

Proposition 3. *The establishment of a transparent system for screening and selection of ideas is associated with the successful implementation of high-involvement service innovation in large firms.*

Our fourth category of managerial antecedents is called "professional innovation employees". Our findings suggested that there is a risk that creative employees will lose their motivation to come up with new service ideas if the firm is unable to assist them in the development of their ideas into new services. New service idea originators are not necessarily the best new service developers, according to our findings. This has arguably received limited attention in the literature, but the complexity of service innovation and the capabilities needed to succeed with service innovation activities (e.g., den Hertog *et al.*, 2010), do indicate that personnel with many different skills need to be involved in service innovation activities. Thus, it makes sense that such complex new service development processes should be coordinated and led by personnel that are trained to do so, or have experience with the management of such activities, especially in large and complex organizations. We suggest Proposition 4:

Proposition 4. *The existence of professional innovation employees that are able to facilitate and lead the development of ideas is associated with the successful implementation of high-involvement service innovation in large firms.*

The four propositions offered serve as tentative recommendations for managers of large firms aiming to increase their firm's service innovativeness through employee involvement. Future research should examine the propositions further, especially by taking contextual factors like for example industrial sector, firm size and cultural characteristics into account.

Acknowledgments

This work was supported by the Research Council of Norway.

Appendix A — Interview Guide for Managers

Questions	Follow-up questions
Please tell about your background and your position in the firm	What type of education do you have? What kind of professional experience do you have? Why did you start in this firm? What type of tasks do you have in the firm? What type of managerial responsibilities do you have? Do you have much contact with customers?
Please tell what you do to make your employees come up with new ideas	What are the sources of innovation ideas in your department? Do your department come up with too many or too few ideas? Are all potential sources of ideas used well? What do you think you should do to exploit ideas better? Do you have any kind of incentives for employees to come up with ideas? Do you have any guidelines for what areas employees should come up with ideas? Are you encouraging your employees to discuss new ideas with each other? Are you encouraging your employees to discuss new ideas with people outside the borders of the firm (e.g., customers/suppliers)? Is it OK that the employees spend working hours identifying/ developing new ideas? How has the implementation of the portal influenced the identification of ideas? What do you do to ensure that employees choose to present ideas they have for your firm?
How did the management work with the idea in the period between it was presented in the portal and the workshop?	Do you think it is smart that ordinary employees may have the opportunity to vote for ideas? Did the firm work on marketing the ideas in other channels than the portal? Did the firm work on development of the idea in the period? If you worked on development; did you work alone or with others? If you worked with others, did you work with people from your firm or with people from other firms? Did the idea originator participate? Did you or your staff spend your working hours on idea development or did you work on it in your leisure time?

(*Continued*)

(*Continued*)

Questions	Follow-up questions
To what degree was participation in the workshop helpful in developing the idea?	Did the group work during the workshop lead to the idea being better specified/concrete? Do you think the groups were composed in a good way? What kind of competence was valuable to have in the groups to be able to develop/clarify the idea? What competence was lacking in the groups to be able to develop/clarify the idea? How dependent was the groupwork on "someone's" preparations? Were all the ideas that the groups worked with devoted much attention, or were there any ideas that got the most attention because of the group composition? Do you think other ideas would have gained attention if the group composition was different?
To what extent do you think the groups were able to select the best idea?	Do you feel that the groups were able to come up with the most valuable idea to present in the plenary session? Were the groups composed in such a way that they were able to say something about the value of the ideas? What information was missing to say something about the value? Should the groups have used a more detailed method when discussing the value of the idea?
To what extent do you think you were able to select the best idea in the plenary session?	Did the participants in the workshop have the necessary knowledge to assess the value of ideas? How much impact did the presentation have on the value assessment? Do you think the result of the poll reflected how good the presentation technique was, or what the value of the innovation idea was?
Please tell what has happened to the idea after the workshop	

Appendix B — Interview Guide for Ordinary Employees

Questions	Follow-up questions
Please tell about your background and your position in the firm	What type of education do you have? What kind of professional experience do you have? Why did you start in this firm? What type of tasks do you have in the firm? Do you have any managerial responsibilities? Do you have much contact with customers?
Please tell about the source of the idea you presented at the workshop	Were you alone to have the idea, or were you a group who had the idea? If you were a group did all the members come from the same firm? Did the firm allow that you spent time on working on the idea during working hours? What is the relationship between the idea and your position in the firm? Would it be possible to come up with the idea without having the position you have?
Why did you present the idea in your firm's portal?	Who was the first you told about the idea? Did you consider other options than presenting this idea for your firm? How much time did you spend on developing the idea before presenting the idea for your firm? Did you get help from others to develop the idea (possibly by whom), before presenting it, or did you do everything alone? Did you only present the idea in the portal, or did you also present it to others?
How did you work with the idea in the period between presenting it in the portal and the workshop?	Did you create a strategy to get as many votes as possible in the portal? Do you think it is smart that other employees may have the opportunity to vote for your idea? Did you work on marketing the idea in other channels than the portal? Did you work on development of the idea in the period? If you worked on development; did you work alone or with others? If you worked with others, did you work with people from your firm or with people from other firms? Did you spend your working hours on idea development or did you work on it in your leisure time?

(*Continued*)

(*Continued*)

Questions	Follow-up questions
To what degree was participation in the workshop helpful in developing the idea?	Did the group work during the workshop lead to the idea being better specified/concrete? Do you think the groups were composed in a good way? What kind of competence was valuable to have in the group to be able to develop/clarify the idea? What competence was lacking in the group to be able to develop/clarify the idea? How dependent was the group work on your preparations? Were all the ideas that the group worked with devoted much attention, or were there any ideas that got the most attention because of the group composition? Do you think other ideas would have gained attention if the group composition was different?
To what extent do you think the group was able to select the best idea?	Do you feel the group was able to come up with the most valuable idea to present in the plenary session? Was the group composed in such a way that it was able to say something about the value of the idea? What information was missing to say something about the value? Should the group have used a more detailed method when discussing the value of the idea?
To what extent do you think you were able to select the best idea in the plenary session?	Did the participants in the workshop have the necessary knowledge to assess the value of ideas? How much impact did the presentation have on the value assessment? Do you think the result of the poll reflected how good the presentation technique was, or what the value of the innovation idea was?
Please tell what has happened to the idea after the workshop	Has the process affected the satisfaction of working in your firm?

References

Aas, T. H. (2017). Managing new service development processes: The case of experiential services. In: K. R. E. Huizingh, S. Conn, M. Torkkeli, S. Schneider, and I. Bitran (eds.), *Proceedings of the 28th ISPIM Innovation Conference*, 18–21 June, Vienna, Austria. Lappeenranta University of Technology Press, Lappeenranta.

Aas, T. H., Breunig, K. J., & Hydle, K. M. (2017). Exploring new Service Portfolio Management. *International Journal of Innovation Management*, 21(6), 1750044.

Aas, T. H., Jentoft, N., & Vasstrøm, M. (2016). Managing innovation of care services: An exploration of Norwegian municipalities. *Cogent Business & Management*, 3(1), 1215762.

Addison, J. T., & Belfield, C. R. (2004). Union voice. *Journal of Labor Research*, 25(4), 563–596.

Bakker, A. B., & Demerouti, E. (2008). Towards a model of work engagement. *Career Development International*, 13(3), 209–223.

Bessant, J., & Caffyn, S. (1997). High-involvement innovation through continuous improvement. *International Journal of Technology Management*, 14(1), 7–28.

Bharadwaj S., & Menon, A. (2000). Making innovation happen in organizations: Individual creativity mechanisms. Organizational creativity mechanisms or both? *Journal of Product Innovation Management*, 17, 424–434.

Cooper, R. G. (2008). Perspective: The stage gate® idea to launch process — Update, what's new, and NexGen systems. *Journal of Product Innovation Management*, 25(3), 213–232.

Dahlander, L., & Gann, D. (2010). How open is innovation? *Research Policy*, 39, 699–709.

de Jong, J. P. J., Bruins, A., Dolfsma, W., & Meijgaard, J. (2003). *Innovation in service firms explored: what, how and why? Strategic study B200205*, EIM Business & Policy Research Zoetermeer, Holland.

de Jong, J., & den Hartog, D. (2010). Measuring innovative work behavior. *Creativity and Innovation Management*, 19(1), 23–36.

de Jong, J. P. J., & Kemp, R. (2003). Determinants of co-workers' innovative behaviour: An investigation into knowledge intensive services. *International Journal of Innovation Management*, 7(2), 189–212.

de Jong, J. P. J., & Vermeulen, P. A. M. (2003). Organizing successful new service development: A literature review. *Management Decision*, 41, 844–858.

den Hertog, P. (2000). Knowledge-intensive business services as co-producers of innovation. *International Journal of Innovation Management*, 4(04), 491–528.

den Hertog, P., van der Aa, W., & de Jong, M. W. (2010). Capabilities for managing service innovation: Towards a conceptual framework. *Journal of Service Management*, 21, 490–514.

Engen, M., & Magnusson, P. (2015). Exploring the role of front-line employees as innovators. *The Service Industries Journal*, 35, 303–324.

Engen, M., & Magnusson, P. (2018). Casting for service innovation: The roles of frontline employees. *Creativity and Innovation Management*, 27(3), 255–269.

Ford, R. C. (2001). Cross-functional structures: A review and integration of matrix organization and project management. *Journal of Management*, 18(2), 267–294.

Froehle, C. M., & Roth, A. V. (2007). A resource-process framework of new service development. *Production and Operations Management*, 16(2), 169–188.

Fuglsang, L., & Sørensen, F. (2011). The balance between bricolage and innovation: Management dilemmas in sustainable public innovation. *The Service Industries Journal*, 31, 581–595.

Gilson, L. L., & Madjar, N. (2011). Radical and incremental creativity: Antecedents and processes. *Psychology of Aesthetics, Creativity, and the Arts*, 5(1), 21–28.

Grönroos, C. (2008). Service logic revisited: Who creates value? And who co-creates? *European Business Review*, 20, 298–314.

Gulati, R. (2007). Silo busting. *Harvard Business Review*, 85(5), 98–108.

Hallgren, E. W. (2009). How to use an innovation audit as a learning tool: A case study of enhancing high-involvement innovation. *Creativity and Innovation Management*, 18(1), 48–58.

Herstein, R., & Mitki, Y. (2008). How El Al airlines transformed its service strategy with employee participation. *Strategy & Leadership*, 36(3), 21–25.

Hipp, C., & Grupp, H. (2005). Innovation in the service sector: The demand for service-specific innovation measurement concepts and typologies. *Research Policy*, 34(4), 517–535.

Huizingh, E. K. R. E. (2011). Open innovation: State of the art and future perspectives. *Technovation*, 31, 2–9.

Jaakkola, E., Meiren, T., Witell, L., Edvardsson, B., Schäfer, A., Reynoso, J., ... & Weitlaner, D. (2017). Does one size fit all? New service development across different types of services. *Journal of Service Management*, 28(2), 329–347.

Kesting, P., & Ulhøi J. P. (2010). Employee-driven innovation: Extending the license to foster innovation. *Management Decision*, 48(1), 65–84.

Lages, C. R., & Piercy, N. F. (2012). Key drivers of frontline employee generation of ideas for customer service improvement. *Journal of Service Research*, 15, 215–230.

Melancon, J. P., Griffith, D. A., Noble, S. M., & Chen, Q. (2010). Synergistic effects of operant knowledge resources. *Journal of Services Marketing*, 24, 400–411.

Melton, H. L., & Hartline, M. D. (2010). Customer and frontline employee influence on new service development performance. *Journal of Service Research*, 13, 411–425.

Ordanini, A., & Parasuraman, A. (2011). Service innovation viewed through a service-dominant logic lens: A conceptual framework and empirical analysis. *Journal of Service Research*, 14, 3–23.

Salunke, S., Weerawardena, J., & McColl-Kennedy, J. R. (2013). Competing through service innovation: The role of bricolage and entrepreneurship in project-oriented firms. *Journal of Business Research*, 66(8), 1085–1097.

Scott, S. G., & Bruce, R. A. (1994). Determinants of innovative behavior: A path model of individual innovation in the workplace. *Academy of Management Journal*, 37(3), 580–607.

Skålén, P., Gummerus, J., von Koskull, C., & Magnusson, P. R. (2015). Exploring value propositions and service innovation: A service-dominant logic study. *Academy of Marketing Science*, 43, 137–158.

Scott, S. G., & Bruce, R. A. (1994). Determinants of innovative behaviour: A path model of individual innovation in the workplace. *Academy of Management Journal*, 37(3), 580–607.

Ulrich, K. T., & Ellison, D. J. (2005). Beyond make-buy: Internalization and integration of design and production. *Production and Operations Management*, 14(3), 315–330.

van de Vrande, V., de Jong J. P. J., & Vanhaverbeke, W. *et al.* (2009). Open innovation in SMEs: Trends, motives and management challenges. *Technovation*, 29(6–7), 423–437.

West, J., & Bogers, M. (2014). Leveraging external sources of innovation: A review of research on open innovation. *Journal of Product Innovation Management*, 31(4), 814–831.

Yuan, F., & Woodman, R. W. (2010). Innovative behavior in the workplace: The role of performance and image outcome expectations. *Academy of Management Journal*, 53, 323–342.

https://doi.org/10.1142/9789811234491_0010

Chapter 10

Needs and Implications of Data in Healthcare-Related Policymaking*

Minna Pikkarainen[†,‡,¶], *Julius Francis Gomes*[§,||], *Marika Iivari*[†,**], *Juha Häikiö*[‡,††] *and Peter Ylén*[‡,‡‡]

[†]*Oulu Business School and Faculty of Medicine, University of Oulu, Pentti Kaiteran katu 1, Linnanmaa, Oulu, Finland*

[‡]*VTT, Technical Research Centre of Finland, Kaitoväylä 1, 90570, Oulu, Finland*

[§]*Oulu Business School, University of Oulu, P.O. Box 4600, 90014 Finland*

[¶]*minna.pikkarainen@oulu.fi*

[||]*Julius.FrancisGomes@oulu.fi*

[**]*marika.iivari@oulu.fi*

[††]*juha.haikio@vtt.fi*

[‡‡]*peter.ylen@vtt.fi*

Abstract. This study explores digitalization in the healthcare sector, and presents insights on the use of data for developing and

*This paper was presented at The XXVIII ISPIM Innovation Conference — Composing the Innovation Symphony, Austria, Vienna on 18–21 June 2017. The publication is available to ISPIM members at www.ispim.org.

organizing preventive healthcare services for young people, which has been identified as one of the most crucial issues to be tackled by public decision-makers. Conducted as a qualitative case study with municipal decision-makers, this study highlights the needs and implications of data as a complex, systemic issue with far reaching long-term impact, where the right kind of data could improve the health and well-being of the society far beyond healthcare domain.

Keywords. Healthcare; data; decision-making; policymaking; public.

1. Introduction

People are more and more embracing a future healthcare system which allows them to get care outside hospital walls, controlling and sharing their health information and user generated content, and receiving improved personalized care. Digitalization has enabled unique kinds of opportunities for generating, collecting, analyzing and sharing data (Leyens *et al.*, 2017). In healthcare, digitalization has been driving advances, e.g., in the availability of biomedical data and genetic makeup, the impact of which has been compared to how the development of ICT changed the society in the previous decades (Horgan *et al.*, 2014). Big data analytics, evolved from business intelligence and decision support systems, can facilitate the analysis of large amounts of data for healthcare organizations "across a wide range of healthcare networks to support evidence-based decision-making and action taking" (Wang *et al.*, 2018, p. 1).

Yet, the systematic use of different "rich" datasets, i.e., big data, personal data, statistical data and social media data, has been limited (Krumholtz, 2014). There are several reasons for this: the data are heterogeneous, and in individual silos; the data are complex, and collected in different ways, and using different techniques; there are real data protection and data governance issues to overcome; no tools exist to make these data accessible to end-users, nor to support the necessary complex analytics for non-experts. These problems result in unnecessary increases in time and costs for patients as well as healthcare service providers (Wang *et al.*, 2018). These issues also make it a huge challenge to link healthcare data to external sources from other public and private sectors. Health data and information is being increasingly transferred across national borders, e.g., through cloud technologies, which adds even more complexity for policymakers, healthcare organizations and the ICT industry (Seddon & Currie, 2013) in determining how to use data for organizing healthcare services.

However, also the timeliness and accuracy of data, i.e., how old the data is, is of relevance for decision-making on healthcare services. Data-based decision-making tools for health policymakers in different levels, e.g., cities, regions and national level have been developed in different European countries, but even at best, they are mostly based on one or two years old authenticated statistical data, that the decision-maker should use to identify current needs and predict future trends related to the specific thematic area. These conditions have resulted in significant challenges for healthcare providers to maximize the uptake of big data, personal data and real-time data platforms.

Most of the attention has been given to improving the value and quality of existing solutions rather than supporting policymaking for creating and developing new services, which has key implications on grasping the opportunities and benefits digitalization has to offer. Despite the complex set of challenges identified above, there are demonstrable benefits of the use of different datasets in health policy and public decision-making. "Rich" data can significant opportunities for researchers, health professionals, and policymakers to "move away from looking at population averages and toward the use of personalized information that has great potential to generate personal, societal, and commercial benefits" (Heitmueller *et al.*, 2014). In healthcare sector, opportunities of the use of rich data include, for instance, real-time tracking of disease/condition outbreaks, predicting of future outbreaks and the development of personalized medical care. Rich and real-time datasets are crucial especially for data-driven decision-making in preventive care, which aims for "making a decision for tomorrow based on today's outcomes" (Sherrod *et al.*, 2010). Moreover, preventive dental care present the most complex and multileveled problem on services, as in addition to the individual, they always touch also the society with long-term implications, hence screaming for new kinds of digital steps. The increase in the amount of and the diversity of data combined with improved storing capabilities and analytical tools offer abundant opportunities to all stakeholders in the healthcare ecosystem (manufacturers, regulators, payers, healthcare providers, decision-makers, researchers) and moreover, data also enables improving general health outcomes when exploited the right way (Leyens *et al.*, 2017). Accordingly, the purpose of this study is to explore how heterogeneous data, e.g., personal data, big data, public data, statistical data, third sector data as well as social media, jointly referred to as "rich data", are currently being utilized in health policymaking in the context of preventive mental care services. The research questions calls "how does the use of data impact preventive healthcare services?" This research question is answered through an empirical study on what decisions are being made to provide and organize such services, and what kind of data is needed.

The paper is structured as follows. First, we will discuss literature on healthcare digitalization and what it means in terms of role of data for organizing healthcare services-related decision-making. Then, we will discuss our methodology and empirical grounding, after which the results are being discussed. The paper ends with conclusions.

2. Digitalization in the Health Sector

2.1. *What is healthcare data?*

Healthcare as a sector historically generates one of the highest amount of data globally in different forms (Raghupathi & Raghupathi, 2014). With the rise of technologies like Internet of Things (IoT), Industrial Internet and forthcoming 5th generation (5G) mobile networks, the prospects are immense for the use of data. Most discussed types of data is "big data", which was coined by Cox and Ellsworth (1997) explaining the visualization of data and challenges it posed for computer systems. Over the period, concepts like

Business Intelligence (BI) emphasized the importance of collection, integration, analysis and presentation of data and how the outcomes can positively impact decision-making (Wang *et al.*, 2018). From there, later in 2009, big data as a concept became revolutionary by proving the applicability across industries. Cottle *et al.* (2013) defined big data as "large volumes of high velocity, complex, and variable data that require advance techniques and technologies to enable the capture, storage, distribution, management and analysis of information". Within the big data literature, "volume, velocity and variety" as 3Vs comes quite frequently to qualify a system to be big data; Raghupathi and Raghupathi (2014) adds "veracity" to make the 4Vs when big data is concerned for healthcare. Veracity refers to "data assurance, that the analytics and outcomes of the big data are credible and error-free (Raghupathi & Raghupathi, 2014). As a concept, big data thus activates the vastly and wildly saturated digital contents used to generate information that helps in knowledge creation (Lohr, 2012).

While big data is assumed to impact health sector positively, the western world still faces challenges related to inadequate integration of multiple systems and poor healthcare (Wang *et al.*, 2018; Bodenheimer, 2005). In the wave of digitalization, healthcare is transforming from a structured-based data (e.g., electronic patient report, diagnosis reports that are formally stored, etc.) towards semi-structured (e.g., home monitoring, tele-health, IoT devices, other sensor-based wireless devices, etc.) and unstructured (e.g., transcribed notes, paper prescriptions, discharge records, digital images, communication messages, radiograph films, MRI, CT images, ultrasound images, videos, etc.) forms of data (Raghupathi & Raghupathi, 2014; Wang *et al.*, 2018). Therefore, as depicted in Figure 1, we choose to extend our observations beyond big data onto the use of various types of data to fully understand the contextual characteristics of the use of data in preventive mental healthcare in particular.

The ever-increasing volume of healthcare data makes more sense to harness the potential of various types of data, which also includes the temporal frequencies of data. These frequencies and relativity of data have significant impact on the decision-making process

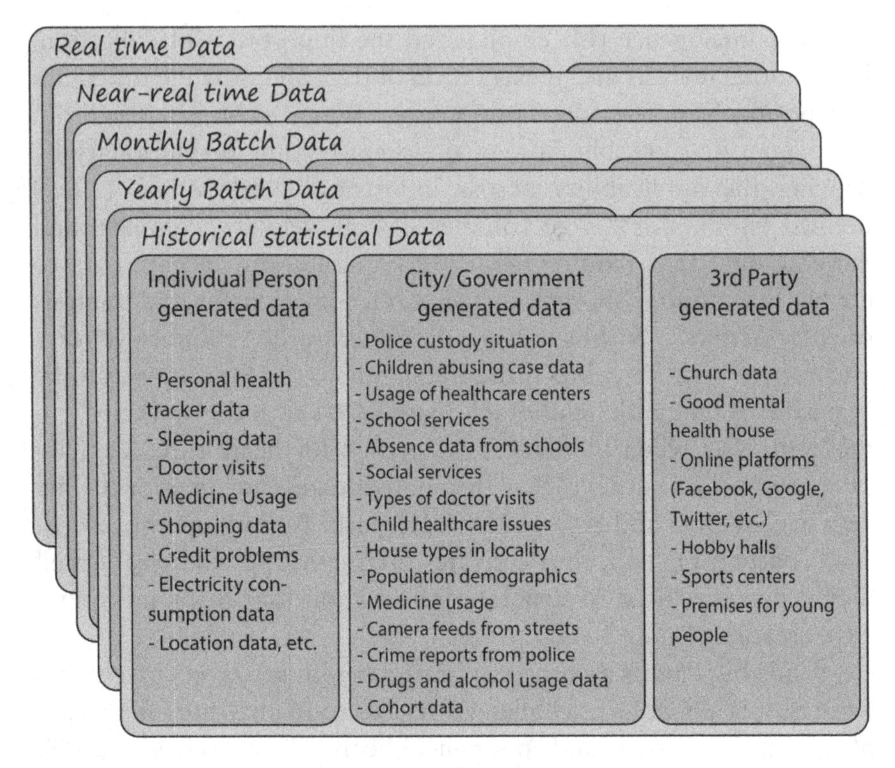

Figure 1. Rich data in healthcare sector.

as well as the outcome. The types of data relevant for organizing mental healthcare services provisioning includes historical statistical data (including cohort data, which refers to monitoring a certain age group over decades), yearly batch data, monthly batch data, near real time data and eventually real time data.

2.2. *Data-driven decision-making*

A sharp drift towards data-driven decision-making is revolutionizing varied fields like science, sports, advertising, education and public health (Lohr, 2012). Various new sources of data are acting as enablers to the effort possible across domains; these data can be structured, semi-structured or unstructured. Exchange and sharing of data between organizations (within the private and public sector is necessary in order to provide integrated governmental services (Otjacques *et al.*, 2014). Decision-making in the public context is about making policies.

Lohr (2012) denotes "big data" as a meme and a marketing term while mentioning the potential of the approach to understand the world of decision-making with completely new dynamics. Speed of data generation is growing enormous in modern time, nearing the global data doubling every 2 years (Lohr, 2012). For the global healthcare sector, mainly flooded with electronic health records the total amount of digitally stored data reached 150 exabytes in 2011 (Raghupathi & Raghupathi, 2014; Cottle *et al.*, 2013). However, the preliminary results from a survey completed by policymakers and public health officials from Finland, Spain, Belgium and the UK indicate that only 57% of respondents agreed that public health policymakers are well informed by public health data before making a decision, and 50% of respondents stated that they only use data sometimes or occasionally "never" for making public health policy decisions. Up to 64% also said that they never perform statistical analysis in making these decisions and 57% admitted that they do not use any simulation or forecasting and often only use census data and data from epidemiological studies. The survey also confirmed that the respondents did not have access to real-time data when making decisions and mostly annual datasets are used. The respondents specified that they would like to have access to diagnostic data and data from similar cities for the purposes of benchmarking. Finally, the respondents have currently ranked the prevention of obesity, mental issues, diabetes, cancer and heart disease (in that order) as the most important areas requiring effective public policies.

Thus, the healthcare sector is still under digitization, i.e., transforming manual and analogical data into a digitized form rather than, digitalization, which refers to a process and use of digital technologies to change current (business) models in order to provide new value generating opportunities (Blake, 2017). Health policies in practice should therefore incorporate both public and private policies. Also, the development and deployment of health policies could take in city/region level, or national (governmental) level, or even cross-national/international level. Policy formulation in general is a complex process according to Brownson *et al.* (2009). Because public health is impacted by numerous determinants outside the health system, when policymakers deal with formulating health policies they should consider those external elements, i.e., scientific, economic, social and political forces (Brownson *et al.*, 2009; Buse *et al.*, 2012). Brownson *et al.* (2009) defined policy to be a broad concept including laws, regulations and judicial decrees and also agency guidelines. More simplistically, Dye (2001) marks what a government do, or do not do, eventually constitutes public policy. Buse *et al.* (2012) further state, public policies can be targeted for specific sector or activity, such as: health, education, economy, and so on. Health policy concerns courses of action and inaction that affect involved institutions, organizations, services, and funding arrangements of the health and healthcare system in place.

Thus, digitalization as a phenomenon should be seen as the enabler of change in the ways in which different domains are restructured in contemporary economy, society and culture (Kreiss & Brennen, 2014). This multi-domain perspective on data is particularly important from preventive healthcare service perspective. Lohr (2012) stated that in future decisions across purposes will be based on data and analysis instead of experience and intuition, especially data aided and analyzed by artificial intelligence techniques. Hence, when we discuss the use and role of data in decisionmaking in the healthcare sector, we can acknowledge that the types of data required for providing, organizing and resourcing healthcare services is multifaceted. The challenge is that, as of now, decision-makers (e.g., public health providers, community level decision-makers, city level decision-makers and governmental level decision-makers), typically do not have any control over the design of the data, its formatting or how it is being collected. The opposite is in fact true; the health providers' data is severely fractured, disjointed, stored in multi-formats and not even always in an electronic format. Data-based decision-making tools for health policymakers in different levels, e.g., cities, regions and national level have been developed in different European countries, but even at best, they are mostly based on one or two years old authenticated statistical data, that the decision-maker should use to identify current needs and predict future trends related to the specific thematic area. These conditions have resulted in significant challenges for healthcare providers to maximize the uptake of big data, personal data and real-time data platforms.

3. Research Design

This study provides empirically grounded implications on how different, complex heterogeneous datasets, e.g., big data, social media data, personal data, statistical data and real-time data are being utilized in municipal and regional levels of decision-making in the context of preventive healthcare, and what implications these decisions have on organizing, providing and resourcing preventive mental healthcare services for the young people.

3.1. *Data collection and analysis*

Data was gathered through semi-structured interviews with key decision-makers to address and identify needs, barriers and future opportunities on data usage in the context of preventive, mental healthcare. Eight representatives from a municipality sector were interviewed in six different interview sessions. Most of the interviewees from a municipality level had also knowledge about wider regional factors, as they were participating in working groups involved in a health, social services and regional government reform. Table 1 presents a group of interviewed decision-makers for this study. Five out of eight interviewees were employees of city of Oulu and three were representing regional perspective, but act as officials in other municipalities due to dual role in regional healthcare reform that is taking place in Finland.

The interviews were all conducted in interviewees' native language Finnish in order to avoid misunderstandings related to specific vocabulary within the research context. All interviews were recorded and then transcribed into English.

Table 1. Interview details.

Role	Date	Duration
• Finance Manager • Development and Quality Manager	February 27, 2017	1:02
• Director of Healthcare and Social Welfare	March 20, 2017	0:53
• Head of Health, Social and Education services	March 22, 2017	1:24
• Director of Social Welfare • Director of Healthcare	March 22, 2017	0:59
• Director of Education	March 28, 2017	1:06
• Director of Joint Municipal Authority	April 10, 2017	1:33

The data were analyzed by the means of thematic analysis (Boyatzis, 1998). This was done to identify the current challenges in regards to data usage, the relevant data sources and types in preventive mental healthcare services, as well as in discovering needs in regards to use of data in decision-making. The set of interview questions is provided as an Appendix A.

4. Results

The following presents the key findings on the use of data for preventive healthcare decision-making. The results are broadly categorized under current use of data in decision-making, use of data in preventive healthcare decision-making, as well as use of data for future decision-making.

4.1. *Current use and needs for decision-making on healthcare services*

How decisions are currently being made looks at what kind of data these decisions are based on in healthcare, also addressing how information is being gathered and produced currently.

Several challenges were identified from service provisioning perspective:

> *"We would need information from several different directions, now we only get it from use of service. That's based on visits ... You get certain kinds of information and certain reports from management system, but that's this "historical data" ... Catching these weak signals and holding on to them ... All kinds of behavior, like shopping gives indications on what to expect. People's hobbies and use of sports facilities. In youth services we follow some of this.*
>
> *We'd need to measure that when we put resources onto light services, how does that show in the heavier ones. Information is at too raw level.*
>
> *The problem is that cause-consequence isn't that clear. Education and culture plus wellness services, why does it show that the red services have increased? There has also been a lot of leukemia and premature birth cases. If we rely too much on raw data, we can draw false conclusions.*
>
> *Integrating research results would be useful. We would need more, like impact evaluations.*
>
> *We have customer data missing in the social services side, quite a lot actually. Like from private clients. Kids, youngsters, families with children, working people ... We have quite a lot of data related to use, but no predictive information really."*

Ideally, the data should allow estimation of parameter values regarding how different factors affect each other. That is, it is not sufficient to simply show point values for these, but the magnitude of dependency should be possible to estimate either based on time series or data on a suitable level of detail, e.g., population level time series of social service use and alcohol use that allows estimation of correlation and delay between the series or alternatively suitable individual level data that shows the strength of correlation between social service use and alcohol use on individual level. In the absence of data, some of the parameters can be estimated from expert opinions (preferably averaging multiple independent assessments from experts familiar with the situation).

4.2. *Use of data in preventive healthcare decision-making*

Organizing resources for preventive mental healthcare services for young people identified substance abuse and unemployment prevention as very close concerns for decision-makers. Thus, the phenomenon is complex and multileveled, and data required for policymaking for such a case will require a wide base of different types of data in the future for visualization. Figure 2 briefly summarizes the different types of data sources that decision-makers are currently using in service and resource allocation situations when they concern the mental well-being of young people. The importance is based on how many times the data source was mentioned by the interviewees.

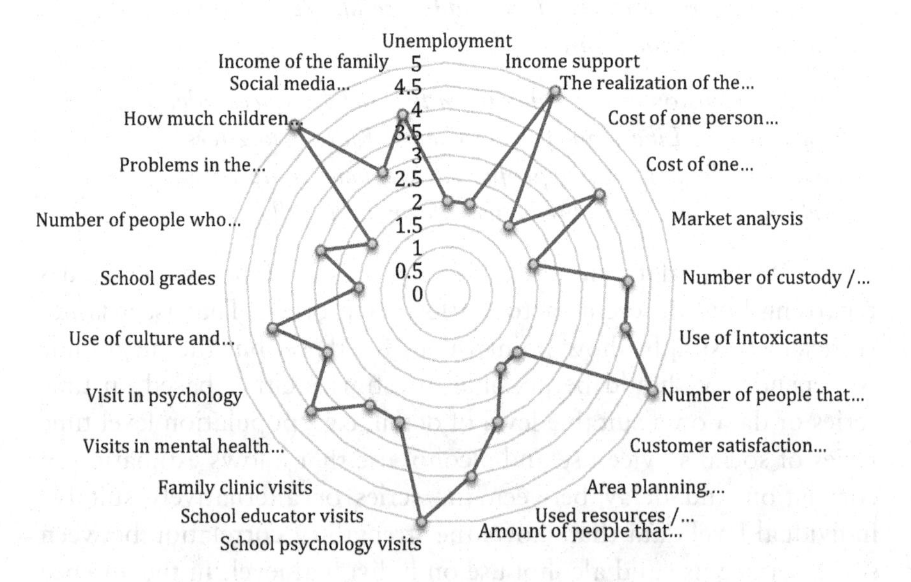

Figure 2. Relevant data sources for preventive mental healthcare.

Table 2. Types and sources relevant to resource allocation.

Data type	Description of the relevant data	Number mentioned
City or Government generated data	Unemployment	2
	Income support	2
	The realization of the economy	5
	Cost of one person working for family support	2
	Cost of one child custody	4
	Market analysis	2
	Number of custody/area	4
	Use of intoxicants	4
	Population	5
	Customer satisfaction surveys	2
	Area planning, environment, e.g., type of houses	2
	Used resources/personnel	3
Third party generated data	Amount of people that stopped vocational education before graduation	4
	School psychology visits	5
	School educator visits	3
	Family clinic visits	3
	Visits in mental health nurse	4
	Visit in general psychologist	3
	Use of culture and education services, e.g., library, church, clubs, etc.	4
	School grades	2
	Number of people who did not get place to study	3
	Problems in kindergarten	2
Individually generated data	How much children move	5
	Social media information	3
	Income of the family	4
	Shopping data	2

The identified data sources were divided in three main categories according to Figure 1: individual person generated data, city and government generated data, and third party (private sector) generated data. Table 2 summarizes these data types and sources relevant to service and resource allocation.

4.2.1. *Data type and source — Individual persons generated data*

Individual and person generated data refers to data produced, utilized and shared by the individual, e.g., through personal health trackers, sleeping monitors, shopping data, personal social media, e.g., chat data, credit problems, location, electricity consumption and so on. From these data types, the movement of young people, shopping behavior and financial situation of the family were mentioned. For the future, interesting additional data sources identified in the interviews could also be personal data produced by children, e.g., data of the child activity level, sleeping, shopping behaviour that can be collected, e.g., from wristbands or wearable devices with the consent of the parents. Typically, this kind of data collection is done through the different campaigns that are established together with Finnish schools and cities.

The personal data was evaluated as a useful data source due to the following key reasons:

> *"Looks like that with youngsters, we have this group that doesn't put effort into studying academic subjects, there should be more of these apprenticeship type education for those who don't read. Those that have been under special needs education, have received better scores, but they fall out when they no longer get support. Those that are worse of normal students, don't get into schools ... There could be developmental delays, behavioural problems, substances, child protection ... This information is divided and with different authorities. What services have been offered to one person, for example ... Has this person gotten a handicap diagnosis later. Have they just surfed through school so that they have never been examined? Now they have been guided to employment services and never received (healthcare) services that they would have been entitled to. These kind of things should be summed up."*

The challenge in the personal data usage is the data protection law that is often not allowing the data usage (at least in the case of under aged people) even the person would give permission into it:

Privacy issues is one thing we need to consider if we want to use personal data. We tried this....

4.2.2. *Data type and source — City/Government generated datasets*

The main data sources were the municipal databases and Finland's National Institute for Health and Welfare (THL) databases. City of Oulu dataset is updated on a monthly basis and includes data from unemployment, income supports, expenses, cost of service workers, cost of child custody, number of child custody, use of intoxicants, population, customer satisfaction surveys, area or environment plans, and resources in use for the services. THL data is collected nationally from healthcare providers like hospitals and cities. THL databases are updated annually and they are integrated based on personal ID. The THL data includes for example hospital register and population-study-based data.

In national level, there is an on-going development to modify legislation related to the use of different data sources. Based on the interviews, the new legislation is constructed in a way, which is compatible to EU data protection regulation. An interesting notion from the interviews was also the problem with scarce data from social services compared to data from health services. The interviewees pointed out that this is an issue which needs development activities, as providing preventive healthcare services to tackle mental health issues are a much wider concern for the society, than, e.g., diabetes. It was said, that if a person has diabetes, that concerns just the individual, but if a person has mental health issues, that concerns also immediate family, relatives, school, work and a person's ability to contribute to society and economy.

4.2.3. *Data type and source — Third party generated data*

Third important data source for the case of preventive mental healthcare services for the young people were the data generated by third party such as schools, church, sport centres, clubs, libraries, premises for young people, private sports facilities, etc. The way that the decision-makers are collaborating with third parties is varying much between the regions, but all of the interviewees acknowledge important information could be gained from membership registers and directly from schools, etc. Furthermore, access to this data is an issue, as so far, legislation prevents the identification of an individual. However, it was stated by the interviewees that in the case of preventive care, that is exactly the need, in order to capture those youngsters before they develop serious conditions or issues with abuse, crime and so on. Active collaboration with schools was highlighted, and schools also have access to collaboration with various kinds of third parties.

4.3. *Use of data for future decision-making*

It was observed that in the future for policy decision-making through simulation and visualization, The most important need was the real-timeliness of data. Currently in many cases statistical data and batch data is being used in form of different kinds of general indicator reports. However, in the future real time data can become more useful in the context of preventive mental health for organizers of services.

> *"The real-timeliness of data matters, in the wellness report there is a lack, since it's updated every two years. We would need information from the ongoing year ... Something could come faster ... Some have data on booklets, some have manual data.. How does information travel between different actors, do we buy child protection and substance services and elderly services. Could be even faster, a substance client needs information right away.*
>
> *"We live in a moment many times, data is more historical information, trends we can look at, but there's no analytical analysis. Data is often old" ...*
>
> *Real-time data? Capacity, situation, daily, customer flows, use of spaces etc. Useful for the organiser who is responsible for*

the entity. Production unit, individual follow-up. Intensive care, customer-centric impact. Long term data ...

The real-timeliness of data at national level is way too old. Economic data should be precise, personnel resource allocation should be precise. Operational data, half a year is maximum. If THL data is from 2015 but that is too old ..."

5. Discussion and Conclusions

The types of decisions healthcare policymakers conduct concern how to allocate resources for certain issues and actions based on data. The challenge is how real-time and accurate the data actually is, as from preventive perspective, public officials should need to be able to react before anything happens. In this regard, certain kinds of data simulations and visualizations would be of immense help, in order to determine outbreaks in certain district/time of year/bigger changes in social structure, e.g., are there more school dismisses in certain region than before, are there more child abuse alerts, custody cases, social worker calls, the implications of unemployment in certain region and so on. Especially data integration from various fields was called for in the interviews in order to track and predict changes over a longer period of time as preventive care also includes services from other sectors, such as education, leisure and culture, or sports and hobbies in general. Preventive care services can be considered as "green services", or situations as usual, whereas interventions, such as custody cases or foster care, indicate severe issues and high alerts as "red services". The goal for providing better preventive green services is that in an ideal situation, red ones would never be needed.

Through exploring how the utilization of different kinds of data, this study contributes to practice by highlighting the complexity of decision-making in the healthcare sector. Digitalization enables faster and better use of data to enabler accurate and tailored services for the needs of individuals. Thus, this study contributes to data-driven decision-making and decision-making theory.

This study also provides public policy implications on the issues decision-makers in national, regional and city levels need to take into account particularly in data-driven healthcare policymaking. All these decision-making levels have direct and indirect implications on individual patients, health professionals, health businesses as well as the society as a whole. The results of the study will be visualized in an on-going MIDAS (Meaningful Integration of Data, Analytics and Services) EU project as concrete solutions that will be collaboratively evaluated with the public decision-makers from different levels identified.

As any study, this also has its limitations. As a qualitative, interview-based case study, the findings cannot naturally be generalized to address the populations of all developed countries, yet alone developing ones. A comparative international cross-case could help to identify further causes and consequences of mental health issues and what kind of data would be most beneficial and useful for policymakers to use. We started off with a clear research plan, however, the complexity of the research problem surprised even us. Therefore, also large scale quantitative studies could largely contribute to solving the challenges policymakers need to deal with today for solving those problems of tomorrow.

Making rich data available for policy decisions is only one preliminary step in the road to data-driven policy decisions. Data analysis and visualization are essential elements in making data usable for decision-makers. Moreover, as traditional data concentrates on current and historical statistics whereas decision-makers increasingly require alternative futures and long-term impacts of the decisions made, there is a definite need for integrating rich data to other tools, such as, foresight, scenario formulation, simulation and sensitivity analysis, to name but a few.

Acknowledgments

The authors would like to thank MIDAS project consortium for their support.

References

Blake, B. (2017). *Healthcare and the Promise of Digitization*. Gartner.Com. https://www.scmworld.com/healthcare-promise-digitization/

Bodenheimer, T. (2005). High and rising health care costs. Part 2: Technologic innovation. *Annals of Internal Medicine*, 142(11), 932–937.

Brownson, R. C., Chriqui, J. F., & Stamatakis, K. A. (2009). Understanding evidence-based public health policy. *American Journal of Public Health*, 99(9), 1576–1583.

Boyatzis, R. E. (1998). *Transforming qualitative information: Thematic analysis and code development*. Sage.

Buse, K., Mays, N., & Walt, G. (2012). *Making health policy*. McGraw-Hill Education (UK).

Cox, M., & Ellsworth, D. (1997, August). Managing big data for scientific visualization. In *ACM Siggraph* 97, 21–38.

Cottle, M., Hoover, W., Kanwal, S., Kohn, M., Strome, T., & Treister, N. W. (2013). Transforming Health Care through Big Data: Strategies for Leveraging Big Data in the Health Care Industry. In *Institute for Health Technology Transformation — iHT*. http://c4fd63cb482ce6861463-bc6183f1c18e748a49b87a25911a0555.r93.cf2.rackcdn.com/iHT2_BigData_2013.pdf

Dye, T. R. (2001). *Top down policymaking*. Chatham House Pub.

Heitmueller, A., Henderson, S., Warburton, W., Elmagarmid, A., Pentland, A., & Darzi, A. (2014). Developing public policy to advance the use of big data in health care. *Health Affairs*, 33(9), 1523–1530.

Horgan, D., Romao, M., Torbett, R., & Brand, A. (2014). European data-driven economy: A lighthouse initiative on personalized medicine. *Health Policy and Technology*, 3, 226–233.

Krumholtz, H. M. (2014). Big data and new knowledge in medicine: The thinking, training, and tools needed for a learning health system. *Health Affairs*, 33(7), 1163–1170.

Lavis, J., Davies, H., Oxman, A., Denis, J. L., Golden-Biddle, K., & Ferlie, E. (2005). Towards systematic reviews that inform health care management and policy-making. *Journal of Health Services Research & Policy*, 10(Suppl 1), 35–48.

Leyens, L., Reumann, M., Malats, N., & Brand, A. (2017). Use of big data for drug development and for public and personal health and care. *Genetic Epidemiology*, 41, 51–60.

Lohr, S. (2012). The age of big data. *New York Times*, 11(2012).

Otjacques, B., Hitzelberger, P., & Feltz, F. (2014). Interoperability of e-government information systems: Issues of identification and data sharing. *Journal of Management Information Systems*, 23(4), 29.

Raghupathi, W., & Raghupathi, V. (2014). Big data analytics in healthcare: Promise and potential. *Health Information Science and Systems*, 2(1), 3.

Seddon, J., & Currie, W. (2013), Cloud computing and trans-border health data: Unpacking U.S. and EU healthcare regulation and compliance. *Health Policy and Technology*, 2, 229–241.

Sherrod, D., McKesson, T., & Mumford, M. (2010). Are you prepared for data-driven decision-making? *Nursing Management*, May, 51–54.

Wang, Y., Kung, L., & Byrd, T.A. (2018). Big data analytics: Understanding its capabilities and potential benefits for healthcare organizations. *Technological Forecasting and Social Change*, 126, 3–13.

Appendix A — Use of Data and Preventive Mental Healthcare Decision-making

INTERVIEW QUESTIONS

Current basis of decision-making

1. How decisions are made at the moment in social and healthcare services?
2. What information are these decisions based on?
3. How information is being collected and produced (tools)?
4. What are the current challenges related to use of data?
5. What kind of data is needed for?
   (A) Allocating resources for regional services? (B) Analyzing the impacts of decisions made?

Use of data in preventive healthcare decision-making

6. What kind of decisions related to preventive mental health, or substance abuse, work?
7. What positive and negative factors impact on a young one's mental health?
8. What kind of short-term and long-term consequences mental problems cause?
9. What kind of decision-making situations you face mostly, when we talk of young people?
10. How services are being produced to support preventive mental healthcare?
11. What kind of collaboration preventive healthcare requires from different actors?

Use of data for future decision-making

12. What kind of data would be needed for enabling better
    (A) Allocation of resources for regional services and (B) Impact analysis on decisions?
13. What is your vision on a world where use of data and decision-making is easy?
14. What factors prevent or enable the use of data?

15. What kind of change is needed for better use of data in decision-making and management?
16. How individual generated data would be used in decision-making (MyData)?
17. Preventive care would be more economical in the long run. How this could be advanced? Is this money off of something else?
18. What kind of opportunities the use of data offers for decision-making and (public) management?

https://doi.org/10.1142/9789811234491_bmatter

Index

Printed in the United States
by Baker & Taylor Publisher Services